To the Mood

'/23

In thanks & apprecia
your home, hospitality, friendship,
and especially your wonderful
children.

With love from the Seattle Jones
family-

Bryan and Janet

1 June 2004

Nathaniel Jones

Alexander Jones   Emily J ll

FORTY YEARS ON ICE

FORTY YEARS ON ICE

A Lifetime of Exploration and Research in the Polar Regions

Charles Swithinbank

Charles Swithinbank
Fulbourn
May 31st 2004

The Book Guild Ltd
Sussex, England

The Book Guild Ltd
25 High Street,
Lewes, Sussex

First published 1998
© Charles Swithinbank, 1998
Set in Times
Typesetting by
Poole Typesetting (Wessex) Ltd
Printed in Singapore
under the supervision of MRM Graphics Ltd, Winslow, Bucks.

A catalogue record for this book is
available from the British Library

ISBN 1 85776 261 4

CONTENTS

Maps

FOREWORD

Forty Years on Ice is a remarkable story. Innumerable books have been written about the polar regions. Usually they tell about one expedition, but here Charles Swithinbank has welded together quite full accounts of his experiences in both the Arctic and the Antarctic. He is a glaciologist who, after gaining invaluable experience with the Norwegian-British-Swedish Antarctic Expedition 1949–52, has been in demand by expedition organizers in many countries.

He has worked with American expeditions, spent 15 months with the Russians in Antarctica, served with the Scott Polar Research Institute, and for many years was Head of Earth Sciences with the British Antarctic Survey. He found himself in the first ice-breaking oil tanker, then under the North Pole in a submarine, and finally as co-pilot on numerous flights in little-known areas of Antarctica.

During these adventures there were many times when rapid decisions were needed – or even inspiration. Charles was never found wanting, and his companions must have been truly thankful for his presence.

So many years, so much achieved – it is good to read about such dedication and the palpable enjoyment that went with it.

Sir Vivian Fuchs, Kt, FRS

PREFACE

My first book about living with ice was *An Alien in Antarctica*. In it I described six expeditions to the New Zealand side of Antarctica. *Forty Years on Ice* is about a dozen different and more varied expeditions, some of them in the Antarctic and others in the Arctic. I considered separating northern hemisphere from southern hemisphere stories but have chosen to keep them in chronological order.

I was first attracted to high latitudes by the challenge of problem-solving in a hostile environment. Too often nowadays, adventure is assumed to be for adventurers – those who choose to travel rough and tough in order to show what is possible or to enhance their self-esteem.

Scientists too have adventures. I retired after a lifetime of research in the polar regions and now feel free to write about the other side of the story. Born with a robust instinct for survival, I never sought adventure but quite often found it. This books tells where and how.

ACKNOWLEDGEMENTS

First, I thank those who gave me the opportunity to take part in these expeditions: The Scott Polar Research Institute, Defence Research Board of Canada, Royal Canadian Navy, British Antarctic Survey, BP Shipping Limited, Royal Navy, Aeroplane and Armament Research Establishment, US National Science Foundation, Fuerza Aérea de Chile and Adventure Network International.

Hundreds of individuals made it all possible. I would acknowledge each of them by name but would live in terror that I might have left someone out. Many friends and colleagues have given continued support over a period of more than 40 years. Without them, I would have no tale to tell.

The following were kind enough to read and comment on drafts of parts of the work: John Ash, Lawson Brigham, Bob Burgess, Chris Doake, Ralph Maybourn, John Noble, Diana Rowley, Graham Rowley, Garry Studd, Mike Walford, Bob Wells, Martyn Williams and Joe Wubbold.

Sir Vivian Fuchs, Geoffrey Hattersley-Smith and Mary Swithinbank read the whole book and did much to improve it.

For photographs and permission to use them I thank the Royal Canadian Navy (pp. 8, 9, 12, 19), US Geological Survey (pp. 35, 124, 173, 184), Ministry of Defence (p. 63), Geoff Somers (p. 138), and Anne Grunow.

Finally, no family could have been more supportive than my wife, Mary, together with Anne, Carol and Kelvin. They encouraged me and kept a happy home throughout my long absences. To them I owe a boundless debt of gratitude.

Charles Swithinbank
Cambridge
December 1996

PROLOGUE (1926–56)

A life sentence to high latitudes

None of my schoolteachers suggested making a career in polar research. It was not on their list because nobody in their ken had ever done it. A few lucky young men had spent two years on an Antarctic expedition before coming home to parents who said: 'Now you must look for a *real* job!' Fewer still went back for a second dose.

How was it that I landed up in such an abnormal occupation? Perhaps my mother was to blame. When I was eight years old she read aloud stories of adventure in far-off lands, of explorers hacking their way through jungle or dragging their sledges across virgin snow. The message she sought to instil was: 'Don't get stuck in an office like your father.'

My father was a classical scholar, educated at Eton and Balliol, who joined the Indian Civil Service and spent his professional life as a District Commissioner in Burma. Burma at the time was part of the British Empire.

My mother's judgment was unjust, because my father did more travelling than any of his contemporaries. But he still spent most of his time in an office.

In Burma of the 1920s, my mother was the only woman who could wield an elephant gun – not for sport but for protection. She was a naturalist who loved animals. She told of swinging her tame leopard by its tail into a tree. Fortunately the leopard too thought this was fun. At the time I was unaware that my mother's love of travel and open spaces must have rubbed off on her children.

In 1933, when I was seven years old, she brought me and my sister Jane to England for schooling. From then on I only saw my father at rare intervals when he was on leave from Burma.

World War II began on 3 September 1939 after German troops invaded Poland. In the same week I went to boarding school – Bryanston – where I spent five happy years. I dreamt – and read – about adventures in far-away places. On three occasions I came close to being expelled from the school: for smoking and drinking, which were against the rules; and for making bombs, which was against the law. The bombs were not for any antisocial purpose – just to enjoy the bang. However, what I learned about explosives proved useful years later when doing seismic sounding in Antarctica.

The family home was just outside Maidstone in Kent. During school holidays I watched dogfights overhead and squadrons of German bombers on

1

their way to blast London. Sometimes they were beaten back by fighters and jettisoned their bombs around us. I delighted in jumping into fresh craters to recover still-warm fragments of the bombs.

At night we sheltered in the family's empty fishpond, which in peacetime doubled as a swimming pool. My mother built a roof over it, added a foot of topsoil and covered the whole with turf to make it look like an innocuous grassy mound. During air raids, shrapnel from anti-aircraft shells often rained from the sky. To protect ourselves on the nightly pilgrimage to the pool, we each held on our head one of the thicker volumes from my father's library.

In June 1940 Jane was evacuated to Canada for safety. But my own baptism of fire brought only a longing to get into the fray and beat hell out of the Germans in any way possible. At school, I joined the Air Training Corps and, after being allowed to take the controls of an aircraft for a few minutes, decided that the Royal Air Force was for me. They told me to wait until I was 18½ years old. By 1944, however, it was becoming obvious that the war might be over before I reached that age. So at age 17½ I left school and joined the navy. That was in the same week as D-Day (6 June 1944) when the allied forces invaded France.

After training as an ordinary seaman, I was promoted to midshipman and served for the next two and a half years in a cruiser, an aircraft carrier and a minesweeper. A visit to Spitsbergen in 1945 left a lasting impression: the spectacular Arctic landscape, the clean air and the pioneering spirit of the settlers made me feel I belonged there. I vowed to come back. The most valuable spin-off of the navy years was a training in navigation that was to stand me in good stead when, years later, I had to navigate everything from dog sledges to aircraft.

Leaving the navy in October 1946 as a sub lieutenant, I went to Oxford to read for a degree in geography. I had no idea where this would lead. During my final year I still had no idea, and to fill in time proposed to join an expedition aiming to map the glaciers of Mount Kenya. Then at an Exploration Club coffee party one day, Scott Russell, a botanist on the teaching staff and an accomplished mountaineer, casually asked if I would like to go to the Antarctic. It took me a fraction of a second to reply. In that moment the course of my life was changed for ever.

Scott Russell had been asked to look out for possible recruits for the planned Norwegian-British-Swedish Antarctic Expedition of 1949–52. They needed an assistant to study ice. Having spent one summer with an undergraduate expedition trying to cross the Vatnajökull ice cap in Iceland, and another with an expedition to the Gambia River in West Africa, I was one of few applicants with an expedition record. From the Iceland excursion I already had a research paper accepted for publication in a scientific journal. But I was only 22 years old and very much a beginner in glaciology – the study of ice. This hardly mattered because I would be working under an

2

experienced Swedish glaciologist, Dr Valter Schytt of the University of Stockholm, and would learn from him.

Psychologists sometimes ask me whether going on an expedition was an escape or a pursuit. I never thought about it. In later years I did meet people on expeditions who were escaping from women, debt, or the law. But for me it was a vocation – or perhaps an unconscious attempt to live up to my mother's admonition to keep out of an office.

In 1949 ours was an international expedition supported by the governments of Norway, Britain and Sweden. The group consisted of six Norwegians, four Swedes, three Britons, a Canadian and an Australian – all of us male. One of the Swedes and three of the Norwegians were married, but I was a bachelor and remained so until 1960. The plan was to spend two years in Dronning Maud Land in the South Atlantic sector of Antarctica. The terrain was unknown and unmapped, though parts of it had been photographed from the air by a German expedition in 1939. Our principal scientific objectives were in the fields of meteorology, geology, glaciology and mapping.

The full story has been well told elsewhere.[1]* The expedition book was published in ten languages; it was so good that in 40 years none of us has tried to better it. On 26 October 1949 I was one of five members of the expedition to sail from Sandefjord, Norway, on board the whaling factory ship *Thorshøvdi*. The expedition's own ship *Norsel* was much smaller and had nowhere to house sledge dogs. *Thorshøvdi* had ample space for our 50 dogs. Besides looking after the dogs, our job was to prepare the expedition's three 'Weasel' tractors for their ordeal by ice.

The rest of the expedition sailed a month later in *Norsel*, a Norwegian sealer. For a month we – the whaling contingent – watched as up to a thousand tonnes[2] of whales were sliced up daily and consigned to boilers below. After 11 weeks we met up with *Norsel* and moved from the big whaler to the small sealer in the lee of an iceberg.

After much searching along the icy coasts of Antarctica, we established our base – Maudheim – on a floating ice shelf that we devoutly hoped was firmly attached to the mainland. Icebergs periodically calve from ice shelves and there was no guarantee that we might not one day find ourselves drifting out to sea. There was no better site, so it was a risk that had to be taken. We unloaded onto the ice everything needed for a two-year stay. This included a third year of food and fuel because there was no certainty that a ship could get through the pack ice every year.

Southwards lay a monotonously level plain reaching to a line 30 kilometres away where the ice surface rose steeply towards a horizon perhaps 500 metres above sea level. Nowhere, not even to the farthest horizon, was

---

* Superscript numbers refer to notes at the end of the book.

there any sign of rock showing above the ice; just whiteness to the east, west and south – and an ocean of pack ice to the north.

Norsel left us after ten days and steamed north. From then on, our little band of 15 lived, if not on terra firma, at least on a thick layer of ice, for two long years. We spent many months exploring inland, mapping, studying the ice and geology of mountains that we found up to 600 kilometres inland. Three lives were lost in a drowning accident; and Alan Reece, one of two geologists, lost an eye after a chip of rock flew into it. The rest of us were lucky. We had narrow escapes but lived to tell about them. Perhaps because of that, and the challenge of overcoming difficulties, this was a most rewarding period of my life.

Two years later, in the first week of January 1952, *Norsel* reappeared to carry us home. This was my last view of Antarctica for eight years. I spent four of these years writing up the scientific results of the expedition for an Oxford doctorate. In due course my work was published as one volume in the expedition's scientific results series.[3]

For the next four years I was based at the Scott Polar Research Institute in Cambridge while studying the distribution of pack ice in the Northwest Passage on behalf of the Defence Research Board of Canada.

Cambridge has been the hub of British polar exploration and research for 77 years. The university oversaw the establishment of the world's first polar institute in 1920. The Scott Polar Research Institute was named for Captain Robert Falcon Scott (1868–1912), who died on the return journey from the South Pole. It houses the world's largest polar library and has staff actively involved in Arctic and Antarctic research. Later, Cambridge became also the headquarters of the British Antarctic Survey (BAS). BAS is a government organization with a staff of more than 400; it operates two ships, five aircraft and three permanent research stations in the Antarctic.

While doing the Arctic work I was invited to return to Antarctica, first by a British, then by an American, and finally by an Australian expedition. But I had signed on for the Arctic, so that was that. Luckily when the Arctic work was completed in 1959, a new opportunity arose for resuming my work in Antarctica.

I was offered a post as Research Associate and Lecturer at the University of Michigan in Ann Arbor, USA. Out of this grew my part in some of the expeditions described in *An Alien in Antarctica*.

In 1960 I acquired an American wife and in 1963 moved back to Cambridge with our young family. From December 1963 to April 1965 I was an exchange scientist with the Ninth Soviet Antarctic expedition at Novolazarevskaya, 800 kilometres east of Maudheim, where I had wintered 14 years earlier.

Since 1965 I have continued to work in Cambridge, first with the Scott Polar Research Institute, then with the British Antarctic Survey, and now independently.

1. ICEBREAKER (1956)

Circumnavigating Baffin Island

I had never been to sea in a *real* icebreaker. Studying the distribution of pack ice in the Northwest Passage gave me the opportunity. I was employed by the Scott Polar Research Institute, though money for my salary came from the Government of Canada. I was to go wherever the job required.

The Distant Early Warning (DEW) line of defence radar stations spread across Arctic Canada was being supplied by convoys of cargo ships. Millions of dollars had been spent on repairing some of the ships after they were damaged by collisions with ice floes. The Canadians wanted an historical analysis of the seasonal and year-to-year variability of the ice in order to schedule voyages at the best times. The work was to involve analysing records found in ship's log books, shore observer's records, diaries, aerial photographs, and ice charts from aerial reconnaissance flights.

In July 1956 I sailed from Liverpool in the liner *Empress of Britain*, disembarking in Montreal a week later. From there I made my way to Ottawa to work at the Defence Research Board, part of the Department of National Defence. My boss was Trevor Harwood, a geologist by training but at 41 years of age an encyclopedia on all things Arctic. On first meeting, he seemed a rough diamond who had little time for the niceties of polite society. Yet for all that, he was one of the most delightful people I have ever worked with. His lively sense of humour buoyed up the whole office. As Head of the Geophysics Section he largely wrote his own brief. A former chairman of the Defence Research Board once remarked that, if he wanted something done in a hurry, he gave it to his Geophysics Section but asked no questions as to how it would be done.[1]

Once I made a travel-expense accounting error that was marginally in my favour. When asked how to salve a guilty conscience, Trevor replied that all civil servants would long ago have perished from remorse if they had consciences.

In Cambridge I had been told that the reason the Canadians had invited me to do this job was because there was nobody in Canada suitably qualified. However, Tom Manning, an old Arctic hand living in Ottawa, said it was because I was doing the work cheaper than any Canadian would. But Trevor had the real answer: 'It is because England is the only country where you could hope to find anyone dull-witted enough to take on such a goddam bitch

5

of a task.' After that judgment I had little to fear from anything he might say behind my back. He was just as unflattering about almost everyone else.

Another person supporting my work was Moira Dunbar, a 38-year-old immigrant from Britain. A geographer by training, she was studying the varied terrains across the vast and largely unexplored reaches of Arctic Canada.[2] She had learned to survive the petty tribulations of working with a man who thought opening a door for a lady was a waste of effort. Once, she told me, Trevor collided with a lady in a doorway. Neither party gave way and they became jammed between door and doorpost.

Then there was Geoffrey Hattersley-Smith, a geologist by training but a glaciologist by inclination. Like me, he had wintered in Antarctica and was a devotee of all things polar. He went on to have a distinguished career studying glaciers, landscape and the history of exploration in high latitudes.[3] Aged 33, Geoffrey too was from England – and never disputed Trevor's judgment about immigrants. Later I learned that Trevor himself was born in Darlington – so here was an office of the Canadian Department of National Defence entirely manned by imports from the old country.

It did not stop there. The head of the Physics Division, of which Geophysics was a part, was Frank Davies, a great Welshman who had been a member of R.E. Byrd's Antarctic Expedition, 1928–30, and leader of the Canadian Second International Polar Year Expedition to Chesterfield Inlet in the Northwest Territories 1932–33. We knew that whatever we did had the full support of the management.

Most of my own work was to be in libraries, but the Canadians wanted me to see some Arctic pack ice in case it was different from the Antarctic ice that I knew. Lieutenant Commander James P. Croal, RCN, made the arrangements for my departure for the Arctic on 17 August. Flying in a Beechcraft to Mont Joli, Quebec, I was put in a Maritime Central Airways DC-4 flying to Frobisher Bay (later named Iqaluit) on Baffin Island. The airline charged not only for me and my baggage but also for the weight of the aircraft seat (22 kilogrammes). On board was Rear Admiral H.S. Rayner, Chief of Naval Personnel, and 20 others heading for DEW-line construction sites. Staying overnight in the VIP quarters (the first and last time I enjoyed that status), I met Captain O.C.S. Robertson, the former commanding officer of HMCS *Labrador*, the icebreaker that I was to join. Under Robertson's command, the ship had completed the first ever deep-draft transit of the Northwest Passage in 1954.

The following day we flew north-west over Foxe Basin to DEW line 'Site 30', also known as 'Hall Beach', or 'Fox'. *Labrador*, I learned, carried two helicopters: a large Piasecki HUP twin-rotor machine, and a much smaller three-seat Bell. The small one awaited us on the tarmac at Site 30. I could see the ship some 5 kilometres away, resting quietly in eight-tenths pack ice.[4] After taking Admiral Rayner, the helicopter returned for me. I think it

Map 1   The Northwest Passage (eastern section)

was my first ride in a helicopter – slightly unnerving but an excellent way to travel over ice that would be impenetrable by a ship's launch.

Unlike most naval vessels, *Labrador* was painted white. I was met on the flight deck by Commander C. Anthony Law, the Executive Officer, who took me to meet the CO. *Labrador* was under the command of Captain Thomas C. Pullen. Like Robertson before him, he had been appointed without previous Arctic or ice-breaking experience. However, I felt somewhat in awe at being ushered into his presence because he had already merited a place in history. Pullen had earned battle honours in the 1939–45 war. When HMCS *Ottawa* was sunk east of Newfoundland, he was one of 70 survivors (114 died). He clung to a crowded life raft for five hours. Many let go, but he hung on. 'I was on my way home to get married,' he explained.[5] Later, at age 26, he was one of the youngest destroyer captains in the North Atlantic.

The icebreaker HMCS *Labrador*

Now 38, he was confident but at the same time showed a disarming modesty. In the Arctic he was carrying on a family tradition, for two of his great-uncles had played prominent rôles as naval officers in the search for HMS *Erebus* and *Terror*, the ships of Sir John Franklin's expedition lost in the Arctic in 1847.[6]

Tony Law seemed an unlikely naval officer but did his job conscientiously and was liked by all. In his heart he was an artist, and in the course of the voyage he painted some stunning landscapes. Out of uniform he could be mistaken for a diffident schoolmaster, so gentle was his manner with people. He took me to the wardroom, where the civilians on board dined with the ship's officers – and had to dress accordingly.

Like most naval vessels, *Labrador* was crowded. The reason must be that one of the design criteria for a fighting ship is that she must be able to function effectively after suffering casualties. It follows that, in the absence of enemies, navies find it hard to come to terms with their own overmanning. We had on board 21 officers, 210 men and 9 civilian scientists, crowded into a ship only 82 metres in length and with 7,000 tonnes' displacement. But I found them cheerful, friendly and cooperative.

The chief scientist on board was Dr Neil Campbell, an oceanographer who shepherded his small flock with a light and easy touch. I had served in the Royal Navy from 1944 to 1946 and in the Antarctic from 1949 to 1952. Now I could enjoy the best of both worlds. Wardroom dress was up to normal naval standards, which, as I was later to find, caused amusement among the Eskimos ashore, where (very) casual Arctic clothing was the norm.[7] On board we changed for dinner. The civilians, having no uniform, were issued with navy blue battledress, only distinguished from the officers by the absence of badges of rank. However, naval officers follow an order of precedence. Thus it was decreed that civilians with a Ph.D. were to be equated with lieutenant commanders, while those with only a BA were lieutenants. All of us wore a white shirt and black bow tie for dinner.

By my standards, the food was superb. Nothing in the diet suggested that we were in the Arctic and thousands of kilometres from any source of supply of fresh meat, fish, poultry, eggs, vegetables, or fruit. Modern navies keep privation at a long arm's length. Drinks in the wardroom were unlimited (within reason) during bar hours and were duty free. The bar was always crowded. There was a laundry service on board, movies twice a week, a small shop, and church services on Sunday.

Captain Pullen briefs the crew

9

Part of the ship's task was to survey as much as possible of Foxe Basin because it was very poorly charted. On 20 August we steamed east to the Spicer Islands. Here the hydrographic survey staff was to take a navigational radio beacon ashore. Most of the islands in this part of Foxe Basin are very low-lying and the sea is shallow. Icebreakers are deep-draft ships – *Labrador*'s draft was 8.5 metres – so we had to proceed with extreme caution. If we had grounded, the nearest tug capable of towing us clear was 3,000 kilometres away.

I was invited to go on a small landing craft that was to establish a depot ashore for the survey work. It was an LCVP (Landing Craft Vehicle and Personnel) weighing 13 tonnes and swung over the side by the ship's crane. In spite of using a lead line to make soundings, we ran aground on a falling tide with no land in sight. It was five hours before we came off on a rising tide. Luckily it was a warm summer day at +5°C, so we were able to keep comfortable by moving about. The Survey Officer finally found a channel to approach the islands and ten seamen unloaded the two-tonne cargo in the space of ten minutes.

The Captain evidently ran a tight ship, but he kept a ready smile and never took himself too seriously. He was not only respected but also liked by his crew. When he learned that I had been in the navy (he made no distinction between RN and RCN), he put me on a regular bridge watch as ice observer. It was a welcome privilege to be made to feel useful. Being here to observe ice, I was in an ideal position to volunteer for helicopter rides to report what lay ahead. This involved sitting in the co-pilot's seat, sketching the distribution, type and concentration of ice to the horizon, and recommending the best route. Meanwhile, whenever the ice permitted, the ship's survey launch *Pogo* zigzagged to make soundings ahead and on either side of our track.

There was a DEW-line supply convoy approaching from the south. Establishing and manning the Distant Early Warning stations was a joint Canadian/US project, so there were to be ships from both countries. *Labrador* would lead one column of ships and the icebreaker USS *Edisto* would escort the other. The pre-arranged rendezvous was off Seahorse Point on Southampton Island at 0900 on 28 August.

I had no concept of the scale of the operation until I saw the fleet approaching from the east. *Edisto* was leading four large military cargo ships, two 'Landing Ships Dock' (LSD), one tanker and a Canadian patrol ship. Some of the LSD craft had been among the hundreds of ships involved in the D-Day landings on the coast of Normandy. The LSD is essentially a self-propelled dry dock carrying landing craft. By flooding ballast tanks to lower the ship in the water, her boats could come in or out of the dock.

It was a grand sight as the fleet moved into line astern behind *Labrador* and *Edisto*. We led USNS *Lieutenant George W. Boyce* (Attack Cargo Ship);

USS *San Marcos* (LSD 25); USS *Fort Mandan* (LSD 21); USNS *Peconic* (tanker); and CGS *Edward Cornwallis*. The whole convoy was headed for Hall Beach, from where some of the supplies were to be distributed by air to less accessible sites.

A convoy proceeds at the speed of the slowest ship. However, when the leader is slowed by ice, everyone astern has to slow down to keep their separation, in this case 500 metres. I was sent up in the small helicopter. We flew ahead for 40 kilometres, then east for 15 kilometres before returning to the ship. The ice covered only one-tenth of the sea as far as the eye could see, so there was no problem. It took the convoy only two days to cover the 600 kilometres to Hall Beach. Our column arrived at midnight to find that *Edisto* was already at anchor, having chosen to lead the faster cargo ships. There was a blaze of lights from the fleet at anchor. *Fort Mandan, Peconic* and *Edward Cornwallis* stayed to unload while the rest of the convoy headed for Rowley Island, a new DEW-line site 100 kilometres to the east.

It soon became clear that this was the largest naval operation in the Arctic since the war. We entertained officers from other ships and they entertained us. The helicopters were kept busy skittering back and forth. I was asked to give a lecture on Antarctica and had a full house – some of the Americans had been to McMurdo station.[8]

Aground, out of sight of land

Labrador spent a week around Rowley Island, landing cargo, surveying in and out among islands and putting up beacons. I saw two beluga ('white whales' of the dolphin family) from the helicopter. We had another grounding adventure in the landing craft while trying to set up a radar reflector beacon for surveyors on Baird Peninsula. Stuck fast 200 metres from shore, we had to unload by manhandling the beacon across wet mudflats. The landing craft was open to the elements and offered no shelter. Wet from wading in the sea, we became extremely cold and suffered fits of shivering. It was 17 hours before our craft floated off on a rising tide.

Since leaving Hall Beach (Site 30) *Labrador* had visited Rowley Island (Site 31), Bray Island (Site 32) and Longstaff Bluff (Site 33) on Baffin Island. Most of these places were windy and desolate and I felt for the men who had to stay for months on end. But they were highly paid, and that was why most of them came north.

I was put ashore at Hall Beach to fly in a US Navy R4D (DC-3) with four other ice observers. One was Walt Wittmann of the US Navy Hydrographic Office, whom I had met before at a sea-ice conference. We flew for six hours, most of it at hedge-hopping height because of low cloud. Heading west, we followed Fury and Hecla Strait, the island-studded and almost uncharted channel that separates Baffin Island from the Canadian mainland. Then north over the Gulf of Boothia, and finally through the narrows of Bellot Strait, which separates the northernmost tip of the North American continent from Somerset Island. Having read some of the history of the long search for the Northwest Passage, for me these place names brought to mind stirring echoes from the past. The landscape looked barren but I knew that it could support foxes, hares, caribou and musk oxen. Polar bears would also be prowling on the ice beneath us.

Our convoy duties were now at an end and Pullen was looking forward to something special. On 16 September, together with *Edisto*, we headed northwest for Fury and Hecla Strait. Our objective was to chart a passage through it and then continue into the Gulf of Boothia. 'At last a break,' wrote Pullen in his diary, 'from the flat monotony of Foxe Basin.'[9] Both ships felt their way through the eastern narrows of the strait – only 2 kilometres wide – and lowered their survey launches when ice permitted. Rather than risking a ship, the procedure in uncharted waters was to send the shallow-draft survey boat *Pogo* zigzagging ahead. But some of the time this was impossible because we were crashing through hard ice floes 3 metres or more in thickness.

Icebreakers have a raked bow, and with it they charge ice floes, riding up and breaking the ice with the weight of the ship. From time to time we came to a grinding halt and backed off. On one occasion Pullen boldly charged, rode up on a heavy ice flow, and stuck fast with the bows high in the air. 'Full astern both!' he called. There was a lot of churning of water but nothing moved. At this point the heeling tanks were brought into play. These are ballast tanks along each side of the ship with massive pumps capable of transferring hundreds of tonnes from one side to the other in less than a minute. Rocking the ship in this way often helps to free her but on this occasion it did not. The solution was to leave both engines going full astern in the hope that the wash would finally weaken the ice floe under the bow. It did, but the stranding – and with it Pullen's embarrassment – lasted a full half-hour. Anxiety had already turned some men's thoughts to the possibility of becoming beset for the winter.

On emerging into the Gulf of Boothia, we became the first ship ever to pass through Fury and Hecla Strait from east to west. It was no mean achievement to have taken a deep-draft ship through a strait characterized by submarine pinnacles. The trickiest part of the passage was later named Labrador Narrows. From here on there was a lot of ice, but it was young (thin) ice that was well within our ice-breaking capacity.

Edisto was lagging far behind us at the eastern entrance to the strait. We now heard over the radio that her starboard propeller had struck a very hard ice floe and shed a blade.[10] She lay helpless in close pack ice. *Edisto* had steel propellers and had lost blades before, whereas *Labrador*, of more modern design, had phosphor-bronze blades which seemed to survive without snapping off.

We had no choice but to go back through the strait and lead her to the relatively open water of Foxe Basin, where she could proceed south using only her port-side propeller.

This done, we began our second westbound transit of the strait. On 20 August we were fighting ten-tenths multi-year floes towards the western end.[11] Out on a reconnaissance flight, our helicopter spotted a small uncharted 'island'. The crew landed on it and were convinced that they had discovered new land. I was suspicious because the ship was in 220 metres of water and they had seen no tide crack. Between any land and the shorefast ice around it there are cracks, owing to the rise and fall of the tide. I asked to have a look at the island. Suspecting that it might only be rocks covering a massive ice floe, I took an ice drill and a lead line. The helicopter pilot left me alone on the ice beside the island and flew off on another errand. After taking a few paces I was horrified to come upon clean and fresh polar bear tracks. Beside the giant footprints was the mandible of a fox. With only a hand auger to defend myself, I felt vulnerable, so rehearsed in my mind the defiant roars that I had been told might deter an attack.

The 'island' measured about 15 by 25 metres and was 4 metres high. The only sign that the rocks, in one case a boulder 2 metres long, might be resting on ice and not land, were small traces that suggested recent slumping. To clinch the matter I broke through the sea ice only 10 metres from the 'shore', lowered the lead line, and found no bottom at 46 metres. Evidently this was an ice floe that had picked up its rocks somewhere else. Through melting from the top in summer and freezing from below in winter, the debris had finally been brought to the surface. The interest to me was that the history of Arctic exploration is speckled with claims to the discovery of islands that were never seen again.

Relieved to be picked up by the helicopter after ten minutes, I reported my findings to the Captain and *Labrador* steamed safely by the 'ice island'. Some days later, having zigzagged up the Gulf of Boothia, we approached Bellot Strait. The Strait was in places only about 1 kilometre wide and the chart showed a single line of surroundings, the result of a pioneering voyage of the shallow-draft RCMP schooner *St Roch* in 1942.[12] With our draft of 8.5 metres, we felt that anything could happen.

It very nearly did. We were being led by the launch *Pogo*. There were no alarms about shallows until an officer on the bridge, peering over the side to look at an area of disturbed water, saw a rock beneath the surface within 20 metres of the ship. Pullen cautiously backed off. *Pogo* had not reported shallow water because there had been none on her track. We had found a submarine pinnacle which was just *off* the track. The only kind of vessel that could have picked up a submerged rock to one side would have been a minesweeper.

The Captain ordered up a helicopter, hopped into the observer's seat and wrote, after looking down on the rock: 'I flew over it and it was a sight to make one's blood congeal. We have named it Magpie Rock – a near miss.' Water was boiling over the rock at 8 knots. After much work *Pogo*'s surveyors found a safe channel 300 metres wide with a minimum depth of 24 metres. Our landing craft visited Fort Ross, a Hudson's Bay Company fur trading post established in 1937 on the north shore of the strait. Although long ago abandoned, the main house was in perfect condition. The door was found locked, presumably as a deterrent to casual visitors or marauding bears.[13]

Heading north for Devon Island, we were coming into historic Pullen territory. Commander (later Vice Admiral) W.J.S. Pullen (1813–87) of HMS *North Star* had spent two winters based at Beechey Island 104 years earlier. His younger brother, Lieutenant T.C. Pullen (1815–88) was second in command. Our own Captain – Thomas Charles Pullen – was named after his great-uncle. He was excited by the association and sought to conjure these ghosts from the past. The details of Arctic history connected with the Pullens were vivid in his mind.

After crossing Lancaster Sound and anchoring in Erebus Bay on the south-west corner of Devon Island, a crowd of us went ashore on Beechey Island. This is where Sir John Franklin's ships *Erebus* and *Terror* spent the winter of 1845–46. The beach was scattered with relics from his stay and also from the many expeditions that later searched for his lost ships. There were barrels and barrel staves all over the place; I found one marked 1853. Hundreds of empty food tins lay about, their lead-solder seals identifying them as belonging to the nineteenth century. Wooden grave markers recorded the burials of three members of Franklin's crew during the winter of

The author on an ice reconnaissance flight

The island that never was

1846. It has been suggested that their deaths may have been due to lead poisoning from food tins.[14] The wooden foundations of two buildings could be made out – one of them built by the crew of *North Star*. It was a desolate place but imbued with history – as much as any other spot in Arctic Canada. Pullen flew in the small helicopter to a bluff overlooking Lancaster Sound and spotted a rock cairn supporting a short wooden mast. His pilot, Lieutenant J.A. MacNeil, found a brass pipe in the cairn. On it was stamped 'RCMP St. Roch 21 Aug. 1944' and inside were notes dating back to 1906. Pullen saw a couple of musk oxen on the adjacent mainland of Devon Island.

We spent 27 September in Resolute Bay, a small Eskimo village on Cornwallis Island built near an airstrip which takes transport aircraft serving the whole of high-latitude Canada. It is the biggest settlement for hundreds of kilometres in any direction. For many of the crew this was their first sight of an Eskimo village, so they were intent on taking photographs. Their preconceptions were shattered when the locals brought out cameras to photograph their visitors. Our helicopter landed among a group of small children playing on a village street. Sensibly, they kept their distance as would any adult; it was obvious that helicopter visits were not unusual.

Eastbound along Lancaster Sound, Pullen despatched a helicopter to the cairn on Beechey Island to correct his misspelling (Beechy) on the note written to record our earlier visit. It is a common mistake but it was evidently embarrassing for one so steeped in the island's history. The place is named after Rear Admiral F.W. Beechey, RN, a leading figure in the Franklin search.

Our next stop was at Arctic Bay, a pretty little Eskimo village on Admiralty Inlet. The government had sacked the radio operator for misbehaviour and wanted to get him out of the settlement as soon as possible. We took him.

The first of October saw us at Pond Inlet, another small settlement. Pond Inlet boasted a policeman, a Catholic mission and a Hudson's Bay Company store. The village faces Bylot Island and enjoys a magnificent view of mountains and glaciers coming down to the sea. Nobody lives on Bylot Island – it is a haven for Arctic birds. The ship's doctor returned from a visit ashore carrying three narwhal tusks bought from local hunters.

Steaming into Baffin Bay, we headed for *Labrador*'s home port of Halifax, Nova Scotia. One thing I learned on this part of the voyage was that an icebreaker, with its hull shaped like a bathtub, can roll to extremes in a storm. The ship's meteorologist measured wind speeds of 100 knots, and a pendulum on the bridge showed some rolls exceeding 45 degrees on either side, meaning that we were rolling through 90 degrees. It paid to hang on to something at all times. Curiously enough, we were aboard the first icebreaker ever built with active roll stabilizers – short wings that project from each side of the hull and alternately turn up and down to counteract the roll. If the

16

Graves of Sir John Franklin's men on Beechey Island

John Torrington's grave from 1846

officer of the watch fails to retract them before the ship gets into ice, he will have a red face and, afterwards, a court martial. Unfortunately, the stabilizers were out of commission because their controlling gyroscope had failed.

It took 12 days to reach Halifax, where I was to leave the ship. En route, we had to tow a cargo ship – *Lady Cecil* – that had broken down and nearly drifted onto rocks. We took her into Corner Brook, a pulp mill in a sheltered Newfoundland fiord. After another day sprucing up the ship we steamed into Halifax on 13 October to be met by TV cameras, newspaper reporters and a crowd of friends and relatives. *Labrador* had been away 103 days and steamed 34,500 kilometres, about 22,000 kilometres of which were in uncharted waters. I had seen a great variety of sea-ice types and now felt confident that I could interpret the records that I was to seek in museums and archives up and down the country and abroad.[15]

Five months later, in March 1957, I was invited to join the ship on a goodwill visit to Norway and Denmark. The purpose of the visit was to 'informally represent all levels of Canadian government in northern activities'. Others invited were F.C. Goulding-Smith, Dominion Hydrographer; Graham Rowley, a senior official in the Department of Northern Affairs and National Resources;[16] and Neil Campbell. We flew the Atlantic in an RCAF aircraft and joined *Labrador* in Portsmouth. Then, crossing the North Sea, the ship and her achievements were fêted at a series of press conferences; we were generously entertained by the Royal Norwegian Navy and Royal Danish Navy; by embassies, polar institutes, universities and libraries. Altogether it was a rewarding public relations exercise and led to increased transatlantic collaboration in Arctic research.

At the end of the visit I was to return to Halifax in the ship. Hoping for a day in Cambridge while *Labrador* was wallowing across the North Sea, I asked Pullen if I could take an airline flight to England and then later be picked up by helicopter as the ship steamed down the English Channel. He said, 'Yes – if the senior helicopter pilot agrees.' On being asked, the pilot said he would try, but on one condition – that I brought him a 3-kilo tin of Fox's Glacier Mints. To this I agreed, and we settled on the time and place of a rendezvous. So I flew to England from Copenhagen.

Days later, I reported at the gate of HMS *Vernon*, the Royal Navy's Torpedo School in Portsmouth, to await my rendezvous. With no paperwork to authenticate my story, the Officer of the Watch looked dubious but accepted that walking me to the helipad – whoever I was – could do no harm. Unfortunately the cloud ceiling was 30 metres above the ground and I thought how silly I would feel if nobody appeared. 'They'll never come in this,' said my escort. At that moment we heard the thump-thump-thump of a helicopter, and minutes later *Labrador*'s huge cigar-shaped twin-rotor helicopter hove into view flying between the twin chimneys of Portsmouth Power Station. 'Cor!' said the officer. By now, I hoped, he was convinced

Polar Bear inspecting *Labrador*

that anyone collected in this way must be a VIP carrying important despatches. He never guessed that my briefcase bulged only with Fox's Glacier Mints.

From the time I returned to Ottawa. I was free to travel to any place that held historical records of the distribution of pack ice in any area between Greenland (in the east) and Icy Cape, Alaska (in the west). It was a fascinating excursion into the history of the Northwest Passage. I had the privilege of meeting or corresponding with some of the great explorers of the century. Vilhjalmur Stefansson gave me his own annotated copy of *The Friendly Arctic* and introduced me to his assistant, Mary Fellows, who later became my wife. My ice work lasted four years and led to the publication of a most unwieldy atlas.[17]

2. ANTARCTICA (1966–67)[1]

Armchair science at 100 knots

I became involved with the British Antarctic Survey in rather unusual circumstances. In 1963 I was working at the University of Michigan in Ann Arbor. For years I had been hoping to learn Russian. The reason was that I had become acutely aware that, collectively, Western glaciologists were ignoring a great deal of Russian published material in my subject simply because of the language barrier. Loathing grammar books, I wanted to learn Russian the easy way, by spending a year as an exchange scientist with the Soviet Antarctic Expedition. Several Americans had done it already and I asked to follow them. However, the State Department in Washington understandably balked at the idea of someone with a British passport being passed off as an American exchange scientist.

In 1959 I had begun corresponding with the Soviet Academy of Sciences in Moscow and the Royal Society in London, but the wheels of diplomacy turned slowly. Another problem was to find an employer who was prepared to do without me for 18 months while at the same time paying a salary into my bank account. It was a cheeky proposition but worth a try.

The Director of the Scott Polar Research Institute (SPRI) in Cambridge was Dr Gordon Robin, an immigrant from Australia and an old friend. We had spent two years together as members of the Norwegian-British-Swedish Antarctic Expedition of 1949–52. Hearing of my ambitions, Gordon secured a post for me at the SPRI on the basis of funding provided by the British Antarctic Survey (BAS). He made clear that on returning from my sojourn with the Soviet expedition, I would be expected to develop a glaciological research programme as a member of SPRI staff. To this I agreed.

So in 1963 I left Michigan for Cambridge – from whence I had come in 1959. The move was not from any kind of dissatisfaction with Michigan; in fact I was very sorry to leave. But the Russians had at last agreed to my joining their expedition, and that, as far as I was concerned, was decisive.

I spent a happy 18 months with the Ninth Soviet Antarctic Expedition, wintering at Novolazarevskaya, and returned to Cambridge in May 1965. I could write a book about the experience, but Gilbert Dewart, a former American exchange scientist, has done a better job than I could.[2]

Back at the SPRI, my chances of taking part in further Antarctic fieldwork depended on collaboration with BAS. Whereas SPRI belonged to the

University of Cambridge, BAS was 100 per cent government-supported and administered by the Colonial Office. BAS had two ships, two aircraft and five research stations in the Antarctic.[3] SPRI's strength lay in its vast polar library and a tradition of research in my field – whereas BAS had specialized chiefly in geology and mapping.

BAS was an entity without a home, though it had its administrative headquarters in London. Responsibilities for research were farmed out to different universities. The head of each unit was employed by the university, but for staff, he had to look to BAS. It was an anomaly, but a workable anomaly. I set about arranging with BAS to recruit graduates who, though employed by BAS, were to work at the SPRI. It was to be ten years before BAS could acquire its own headquarters to bring together the administrative and scientific units.

BAS had a curious origin. Britain's permanent presence in Antarctica dates back to 1943 when Operation Tabarin was established to provide military surveillance and meteorological information for the South Atlantic. The operation was secret, but after the war it evolved into a purely civilian research enterprise known as the Falkland Islands Dependencies Survey. This led to Antarctic personnel being known as 'Fids'. The epithet persists to this day, although the organization changed its name in 1962 to British Antarctic Survey. Those who have moved on to other employment are 'ex-Fids'.

Most Fids were recruited in their early twenties and had come into the organization with little or no outside experience. I was considered an oddball in that I was 39 years old, had never served at a BAS base, and anyway worked for SPRI. The fact that I had three Antarctic winters and eight summer seasons under my belt did not seem to count because they had been with *foreign* expeditions. BAS at the time was a decidedly – and parochially – British organization, believing that it had little to learn from outside. This was inherited from its wartime origins, when international relations were the last thing it wanted.

In 1965, too many glaciologists had been ignoring the third dimension of glaciers – the thickness of ice – because there was no easy way to measure it. Some years earlier I had met an American, Amory Waite, who had served as radio technician on the Second Byrd Antarctic Expedition at 'Little America' in 1934. Waite noticed that Very High Frequency radio communication was possible *through* an iceberg, although physicists had reported that ice was opaque at radio frequencies. It was 24 years before Waite had an opportunity to take the matter further. In 1958 he made depth soundings through a glacier from a sledge and in 1960 he showed that radio echoes, or reflections, could be obtained from an aircraft flying over the ice. Part of the signal was reflected from the surface and part from the glacier bed. The time difference between the two reflections gave a measure of the ice thickness. Waite's instrument was not designed for the purpose, so could only get reflections from the bed where the ice was less than about 380 metres thick. Unfortunately, we already knew from seismic soundings that most of the Antarctic ice sheet was much thicker.

Waite was employed by the US Army Signal Corps, which was more interested in developing aircraft radar altimeters that would *not* penetrate ice. So Gordon Robin decided to build a pulsed radar designed to measure the greatest possible ice thicknesses. The designer was Dr Stan Evans and his 1963 prototype was mounted on a tractor and tested in the Antarctic by Mike Walford.[4] Evans saw no reason why it should not work equally well from aircraft.

Geoffrey Hattersley-Smith, still with the Defence Research Board of Canada, secured the use of a DHC-3 'Otter' aircraft to test the concept over Ellesmere Island in Arctic Canada. Robin and Evans brought along their instrument and installed it in the Otter. The trials took place in April 1966 and were eminently successful, recording ice depths of up to 660 metres.[5] A film camera had been modified so that film moved slowly across the face of an oscilloscope display graduated in microseconds. The effect was to trace a continuous cross-section of the glaciers over which they flew. In subsequent developments the technique has variously been known as radio-echo sounding, radar sounding, electromagnetic sounding, or simply radioglaciology.[6]

After the Arctic trials, the logical progression was to test the instrument in the Antarctic. So eight months after the Ellesmere Island trials David Petrie and I took the instrument south. David, 26 years old, had already served two years at Halley Bay in the Antarctic as an ionospheric technician.

I joined the BAS ship RRS *John Biscoe* at Southampton on 24 October 1966.[7] The Master, Tom Woodfield, was a mild-mannered and somewhat introspective mariner to whom I took an immediate liking. He quickly made me feel at home in the wardroom. As senior scientist on board, I was consigned in solitary state to the 'Governor's' cabin, where Prince Philip, Duke of Edinburgh, had lived during his visit to Antarctica in 1957.

Biscoe was 67 metres long and powered by diesel engines developing 1800 shaft horsepower. I had chosen to travel all the way south by ship rather than the alternative of flying part-way because it was my first field season with BAS and I felt a need to get to know the organization from the inside. Not everything can be learned by discussions in an office.

Petrie joined the ship three weeks later in Montevideo, Uruguay. Stopping in the Falkland Islands to pick up some cargo, I lunched with Sir Cosmo Haskard, the Governor, and Lady Haskard. The tenants of Government House in Stanley took a strong interest both in BAS and in the politics of Antarctica. There had been a historic development in the political status of the continent since I first came south in 1949. A treaty was signed in Washington, DC, in 1959 guaranteeing free access to the continent for any peaceful purpose. The signatories of The Antarctic Treaty, as it is known,

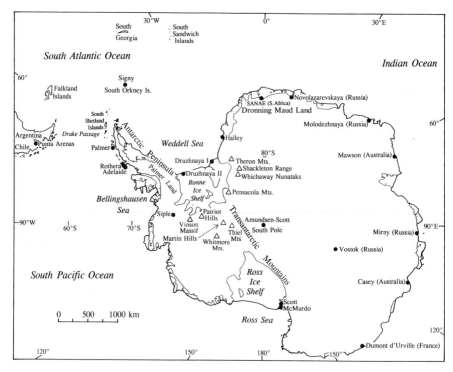

Map 2   Antarctica

were the governments of Argentina, Australia, Belgium, Chile, France, Japan, New Zealand, Norway, South Africa, the Soviet Union, the United Kingdom and the United States. Each of these countries had recent experience of collaborating in the cause of science during the International Geophysical Year of 1957–58. Harmonious relationships were continuing despite the stresses of the Cold War elsewhere.

The need for a treaty arose from a common interest in preventing the militarization of Antarctica, and the need to accommodate differing views on sovereignty. Seven of the signatory nations had made claims to sectors of Antarctica: the British, Argentine and Chilean sectors overlapped. None of the claims was recognized by the United States. Reconciling these differences was a brilliant feat of diplomacy, and the treaty laid the groundwork for cooperation that has stood the test of time.

John Biscoe left Stanley on 28 November and met pack ice two days later. Our first stop was Signy Island in the South Orkney Islands. Signy was a BAS biological research station for which we carried cargo. After landing the cargo and mail we steamed west, and on 9 December anchored off the BAS station on Deception Island. Deception is a sunken volcanic caldera

24

with a harbour inside it. Eleven men had passed the winter here.

Three days later we were at the American Palmer station on Anvers Island. Our plan was to disembark at Palmer to wait for a BAS aircraft to fetch us. The only BAS aircraft in working order at the time was a single-engined Otter. Today, flying anywhere in Antarctica without a back-up air-craft would be regarded as reckless, but in 1966 it was considered a neces sary risk. We were there to get on with the job, not to find reasons for sitting on the ground.

Palmer was a hive of activity because a new station was being built 1 kilo-metre from the first, itself built beside an abandoned BAS station. The Americans looked after us generously while we waited for flying weather. It took the Otter three flights to carry our equipment plus some other urgent cargo from Palmer to Adelaide – the BAS station at the southern extremity of Adelaide Island. The distance was only 430 kilometres but the weather en route was often poor.

The aircraft was fitted with combined ski-wheel landing gear. Here we were operating off snow, so the skis were set below the wheels. The pilot was Flight Lieutenant Robert Burgess. In normal life an RAF Transport Command pilot, Bob had spent the previous winter at Deception Island. He was good company and a good pilot. With us was Bob Vere, a 29-year-old RAF pilot coming south for the first time, and David Petrie. Not long before, Vere had been flying four-engined Shackletons on maritime reconnaissance, so the Otter was a bit of a comedown.

Although I had spent years in Antarctica, this was my first view of Graham Land – the backbone of the Antarctica Peninsula. It consists of a long, narrow, flat-topped plateau rising 1,500–2,000 metres above sea level. Geologically, it is an extension of the Andes of South America. Both its Atlantic (eastern) and Pacific (western) sides are dissected by valley glaci-ers cascading in steep icefalls to the sea. The west coast has an archipelago of islands, some with sharp alpine peaks, others with dome-shaped ice caps reminiscent of icing on a cake. From our vantage point, the view was spec-tacular, and I quietly thrilled at the privilege – once again – of enjoying an Antarctic mountain landscape.

After two hours of feasting our eyes on this ever-changing scene, we land-ed at Adelaide. The airstrip,[8] 60 metres above sea level on unprepared snow, was somewhat alarmingly situated between conspicuously crevassed slopes leading down to a dark blue sea. We were met by a team of tail-wagging huskies and Roy Brand, the aircraft engineer, who at 54 years of age was the old man of the station and 14 years my senior. A tractor drove us down to the station, built on rocky outcrops beside the sea a kilometre from the airstrip.

Seven men had wintered at Adelaide under the leadership of George Green, a diesel and tractor mechanic. The arrival of six fresh faces was something of a shock to the integrated family atmosphere. It was as if we

Map 3   The Antarctic Peninsula

26

had wakened the winterers from hibernation. Some clung to it, sleeping through breakfast and rising only at lunchtime. David and I had a hard time chivvying them into action. Ours was to be a short field season and we could not afford to waste time.

Although there have been as many research stations in the Antarctic Peninsula area as in the whole of the rest of the continent, the region remains relatively unexplored in glaciological terms. The reasons are not so much human or even political as they are environmental. Much of the terrain consists of precipitous mountain glaciers plunging from alpine peaks or plateaux into an ice-choked sea, of calving ice cliffs and hanging glaciers, of steep ice piedmonts furrowed by crevasses, of rugged massifs fringed by icefalls, and an archipelago of ice-capped islands. Research stations of necessity cling to narrow footholds of uneven rock between encroaching glaciers and the cold sea. Adelaide was one such station.

Bad weather is all too prevalent. The peninsula serves as a climatic barrier between its Pacific coast and its Atlantic coast. Persistent low cloud, fog, snowfall, and even rain in summer are accompanied by shifting and turbulent winds.

On 22 December Peter Bird, the station meteorologist, recorded a wind averaging 75 knots for 10 minutes, with gusts to 93 knots. The buildings strained against their anchor cables and we were concerned for the Otter at the airstrip. But it survived. A day flight to Stonington station on Christmas Eve rekindled my perennial fear of crevasses. We were driven from their airstrip to the station sitting on a dog sledge that left many a black hole when snow bridges collapsed under the runners. I chose to use skis for the return journey to spread my weight and span the crevasses. The men at the station had already learned harsh lessons about the hazards of travel in Antarctica. Eight months earlier, a search party from the station had found the frozen bodies of two of their colleagues outside a bivouac. The men – together with every one of their sledge dogs – had succumbed in a ferocious blizzard.

Because of this tragic event, Christmas at Adelaide was a very low-key affair. We slept late, decorated the dining room, and then opened the presents from home. The feast consisted of:

> Lobster tails
> Russian crab
> Filet steak
> Diced ham
> Potato chips
> Sweet corn
> Anchovies
> Christmas pudding with rum sauce
> Meringue pie

Beer, port and liqueurs
Blue cheese

By 27 December David Petrie had installed and tuned the radio-echo sounder in the Otter and we were ready for a test flight. All take-offs were downhill towards the sea, so an engine failure on take-off would inevitably lead to a swim and most likely to drowning. There was no profit in thinking about it. On this flight we were airborne for nearly two hours, crossing to the mainland and flying up and down Martin Glacier. Burgess flew us on the outward leg, changing halfway so that Bob Vere could get the feel of the Otter. David spent most of the time peering into his oscilloscope with a worried look on his face.

On developing his recording film back at Adelaide, we found it was blank. So it was back to the drawing board. Some days were lost to weather, others to domestic duties. A BAS tradition – a good one in the circumstances – is that everyone, whatever his status, takes a turn at washing, tidying, sweeping and providing water for the kitchen. Rank is next to irrelevant in a community of interdependent individuals.

Our second flight brought frayed nerves. In the climb out over the cold sea, suddenly black oil sprayed all over the windscreen, engine cowl, struts and wing roots. We could scarcely see out. It was no place to be with a single-engined aircraft that would sink fast if it came down. Burgess executed a hasty 180-degree turn, while each of us wondered how long the propeller would keep turning. I crossed my fingers and prayed, but we made the land while still under power. Roy Brand stemmed the oil leak, luckily finding no engine damage.

David had done much to improve the radio-echo sounder. On the last day of 1966 we celebrated by getting really good ice-bottom reflections over glaciers on the mainland. We had been warned about severe turbulence over the glaciers flowing down from the plateau and now we were to experience it first-hand. There were moments when I feared for the structural integrity of the aircraft. It creaked and groaned but we lived to tell the tale. Three months later we were to discover how close we had come to disaster.

With the instrument now performing as it should, we flew over ice shelves, valley glaciers, and the plateau, acquiring hundreds of kilometres of good cross-sections of the landscape. I thought back to the days, 15 years earlier, when Gordon Robin and I used seismic soundings to find the depth of ice. After a hard day's work drilling a shot hole, setting out seismometers, detonating TNT and developing the film, with luck we would have achieved a single measure of ice thickness at one point only. Then we had to drive our Weasel tractor at a speed of four knots to the next site. Now in 1967 we were travelling at 100 knots in comfortable seats, with the radar measuring ice thickness thousands of times per second and yielding a nearly unbroken

28

cross-section of the ice sheet.

All the time we were flying, I had to record exactly where we were so that we could reconstruct the flight tracks afterwards and plot these on charts. This sometimes involved making 30 or more entries in a notebook for each hour of flight. My only credentials for navigation came from a three-week course at the Royal Naval College at Greenwich when I was a naval officer in 1945. Later I served as navigator in minesweeping operations. That was an excellent background for navigating any kind of vehicle, because the consequences of a miscalculation might be explosive in a literal sense. I learned to check my arithmetic. Although things happen faster in the air than at sea, the principles are the same.

One of the sadder chapters of Antarctic history will record that conventional mapping, instead of anticipating a predictable need, has in practice lagged far behind. Indeed, the far side of the moon is better mapped than parts of our own planet. Nearly all the maps available to us were sketch maps with errors in places amounting to many kilometres.

This season our objective was exploratory sounding. Perhaps later seasons could concentrate on experiments to throw light on specific unsolved problems in glaciology.

The aircraft had a lot of survival equipment to ensure that if we did make a forced landing, there would be no panic reaction nor immediate request for assistance from foreign expeditions. Both Argentina and Chile had aircraft operating within 800 kilometres. We each carried our own sleeping bag and also extra clothing, while the Otter itself was equipped with:

	kg
3 rucksacks containing food	46
3 winter sleeping bags	29
1 inflatable dinghy	28
1 tent	11
3 pairs skis	9
1 stove, fuel, utensils	6
3 exposure suits	6
3 life vests	6
3 ice axes	5
1 man-haul sledge	4
3 pairs crampons	1
Total weight (kg)	151

Adelaide was on an island, separated by more than 150 kilometres of sea from the areas that I most wanted to visit, so we had to live with the possibility that, one day, there might be an engine failure over the sea. A high-wing monoplane like the Otter is the worst kind of aircraft for ditching. Landing on water, it sinks to the wings and quickly floods the cabin. We doubted that the inflatable dinghy could ever be launched fast enough. With luck, we might choose between drowning fast or dying slowly from hypothermia in the dinghy. Either way, our colleagues on shore had no means of rescuing us. Though aviators must live as optimists, the potential for disaster was always at the back of our minds.

On 3 January we had another little adventure. The Otter had several other duties besides glacier sounding. Various sledging parties with one or two dog teams were at work on the mainland. Each party consisted of a geologist or topographic surveyor with a field assistant (GA or general assistant in BAS terms). They had left the main sledging base at Stonington Island, 100 kilometres south-east of Adelaide, some weeks earlier and were now far afield. At monthly intervals we had to take them dog food, man food and Primus stove fuel, returning to base with the rocks that the geologists had collected in the course of their research.

We were searching for one of the sledging parties comprising a surveyor and his GA working on the Palmer Land plateau. The name refers to the wider part of the Antarctic Peninsula south of Graham Land. The ground party told us over the radio that they were camped somewhere among the Eland Mountains. Bucking a headwind along the Eternity Range, we passed level with its three gleaming white peaks – named Faith, Hope and Charity by Lincoln Ellsworth, an American airman who discovered the range in 1935.

As we flew on, the snow surface became featureless. We were in white-out, a condition in which daylight is diffused by multiple reflection between the snow surface and an overcast sky. Contrasts vanish and there may be no horizon. My diary says, 'The map bore no relation to the ground so map reading was useless.' We spent 40 minutes searching among small rock outcrops that, here and there, jutted above the ice sheet. Eventually the campers heard the plane and guided us by radio towards them. Bob Burgess was in the pilot's seat and I was beside him, while Bob Vere was in the cabin behind. The ground party, Dick Boulding and John Noble, had marked a landing strip with a line of food boxes, and recommended that we land 'towards the tent'. For reasons I did not understand at the time, the Otter touched down halfway along the strip and 300 metres short of a small peak with an ominously long snow tail extending across our path. On our side of the tail there was a windscoop, a chasm sculpted by wind. I have seen windscoops 30 metres deep and several kilometres in length.

At this point Bob must have regretted that ski aircraft have no brakes. The plane slid on with little sign of slowing. We were powerless to stop before the windscoop. Finally, he opened the throttle, attempting to get back into the air. But the machine accelerated slowly because we were 1,500 metres above sea level. Topping a rise, we saw to our horror that ahead, the surface

King Penguins marching

Icebergs and ice floes

RRS *John Biscoe* at Signy Island

Adelaide station. The airstrip is up the ice slope to the left

The Otter with David Petrie (holding radio-echo aerial), Bob Vere, and the author

Fossil Bluff (centre) seen from George VI Sound. The curious round feature has not been explained

Fossil Bluff hut with dog sledges on the roof

David Petrie with his radio-echo sounder in the Pilatus Porter. The yellow inflatable dinghy was intended as a placebo in case we ditched in the sea

fell away steeply. We were nearly airborne but not quite. Careering down the steep slope, I put both hands against the front panel to cushion the expected impact. At some point the slope abruptly changed from down to up. Compressed into our seats by the switchback, we were catapulted into the air with the stall-warning reminding us that our airspeed was precarious. Both of us were shaken by the experience.

The ground party's instructions had led us to land downwind and downhill instead of upwind and uphill, and in the near whiteout conditions, the pilot had not understood this before it was too late. There was no difficulty landing in the opposite direction. Stepping onto the snow, Bob Vere was too polite to say what he thought of the first landing. I have noted over the years that pilots refrain from gloating over another pilot's problems – they know that next time the boot may be on the other foot. I cuddled each of the sledge dogs in turn – more for my own comfort than for theirs. The four of us then unloaded the supplies and flew home. It had been a close call.

3. PLUMBING THE DEPTHS (1966–67 continued)

The rubber dinghy – our placebo

Some radio-echo flights were successful, others not. The aerials were clamped to the wing struts; we moved them from time to time in an attempt to fine-tune the system. On one clear day we were flying north along the peninsula plateau. In the distance to our right we could see to the Atlantic (Weddell Sea) coast, and to our left the Pacific. We were to resupply two field parties. Bob Vere was in the left-hand seat. When the flaps were down and we were gliding in for the first landing, I saw that he was unwittingly heading for a field of snow-bridged crevasses. New to the Antarctic, he had not recognized the faint parallel lines in the snow. I had to warn him, but there was no intercom system in the aircraft and he was wearing headphones to keep out the engine noise. There was only one way to act fast. I picked up the radio microphone and said, 'I advise you not to land here.' Bob reacted quickly and pushed the throttle to the firewall. However, in keying the microphone to speak, I had unavoidably broadcast to anyone – anywhere – who happened to be tuned in. We learned later that two station radio operators had logged this cryptic message without knowing from whence it came.

The next landing was on Cole Glacier. This time the geologists had camped in a crevassed area without realizing it. They expected us to land beside them and seemed upset at having to ski 2 kilometres to the safe spot where we chose to put down. Often enough in earlier years, I had been the one on the ground. Knowing that airmen generally spot snow-bridged crevasses that a sledge party fails to see, we used to long for a kite, or a balloon, or a microlight aircraft to get a better view. Sledge travellers cross many thousands of crevasses that they would have avoided if they had known about them.

On 14 January we flew eastwards over the plateau and right across Larsen Ice Shelf to the Weddell Sea. It was my first view of the kind of heavy pack ice that – not far from here – crushed and sank Shackleton's ship *Endurance* on 27 October 1915. His crew had to live on the shifting ice floes and drag their boats over them, but they survived and ultimately escaped without loss of life.[1] We looked down on the same chaos of hummocks and ridges and knew how lucky we were. This flight yielded our first good cross-sections of Larsen Ice Shelf. Close to the peninsula land mass the floating ice was up to 500 metres thick; from there it tapered off eastwards to 150 metres at the Weddell Sea ice front. It was our longest flight yet – four hours in the air.

By this time, the new BAS aircraft, a Swiss-built Pilatus 'Porter', had been put ashore at Deception Island with its pilot, Flight Lieutenant John Ayers, RAF. It had come south by ship with the wings detached and now had to be assembled. Like the Otter, it was single-engined; but unlike the Otter, it had a gas turbine engine and represented BAS's first foray into modernity. From now on the drawback, while we kept the Otter, was that separate supplies of aviation gasoline (avgas) and aviation turbine fuel (avtur) had to be maintained.[2] We waited anxiously for the arrival of this new machine, knowing that at last a back-up aircraft was available to rescue us in case we made a forced landing.

Flying east on 26 January on top of cloud with the object of making further sections across Larsen Ice Shelf, we began an instrument let-down from an altitude of 2,400 metres when my dead-reckoning navigation indicated that we were clear of high ground. Descending to 900 metres through unbroken cloud, we realized that there was no safe way to get below it, so began to climb back up. Unfortunately we had been picking up ice on the wings, prop, flaps, struts and aerials. The heavier our load of ice, the slower our rate of climb, and many anxious minutes passed before we broke into sunshine on top of the cloud. The ice melted and we quietly resolved never to get into that situation again.

The Porter arrived later the same day, having flown from Deception Island in three and a half hours, the last two hours without sight of the ground. We knew that Sir Vivian Fuchs, the Director of BAS, strongly discouraged instrument flying but Ayers was anxious to demonstrate his skills. By the time he arrived overhead, it was too late to do anything else. So he continued south for a few kilometres after passing our radio beacon, knowing from the map that he would be over the sea, and started down. We saw his ice-covered aircraft appear out of the cloud a few hundred metres above the waves. On landing, he came to a dead stop in 50 metres – BAS now had its first-ever aircraft capable of reverse thrust.

John Ayers lived for flying and talked about little else. He had served as a Farnborough test pilot in jet fighters and could tell stories lurid enough to put anyone off flying. One of his jobs had been taxiing jet aircraft at high speed into a gravel bed to see what would happen if the brakes failed. In Kenya, he had once flown a Mau-Mau prisoner and took pleasure in scaring the man witless by doing aerobatics. It was the Kenyan's first flight and probably – if he could choose – his last. In the Antarctic, John was determined to fly although he would have to use avgas instead of the avtur fuel for which his engine was designed. The first supplies of avtur were due to arrive later in the season in *John Biscoe*.

On 1 February I flew with John to Fossil Bluff, 400 kilometres south on Alexander Island. Over the island he showed off the versatility of his machine by gently applying reverse thrust in the air. This allowed a quite alarming dive angle – steeper than an angry Stuka.

Southward view of George VI Sound
Palmer Land (left) and Alexander Island (right)

35

Fossil Bluff was the loneliest outpost in the BAS inventory. A strongly-built, one-room hut had been flown here in sections some years before. At intervals it had been used year-round for four-man geological parties. The same room served as kitchen, dining room, bunk room, recreation room, meteorological office and bathroom. It had always offered a stern test of human compatibility during the months of winter darkness. Now in summer it was manned by two meteorologists and served as an advance weather station for the aircraft. Windows faced towards the mainland across George VI Sound, giving a panoramic view of big glaciers tumbling down from the plateau. Peter Bird came here by choice for the summer, describing it as an ideal spot for a honeymoon. Some years later he did marry but never managed to bring his bride this far south.

On saying our farewells, Ayers strapped himself in and pressed the starter. Three attempts to start the engine failed. So this was the price to be paid for using the wrong fuel. We spent the night with the natives, though with only four bunks in the hut, some of us slept in tents outside. Bob Vere arrived with the Otter and its engineer the following day. With expert help, the Porter was soon flying again.

The main task of everyone at this time was to restock Fossil Bluff for the winter. Both aircraft spent many hours ferrying cargoes south. Six flights, each carrying nearly a tonne, could be made in a good day. At this rate, the pilots were able to finish the job in only three days. Having completed the resupply, we were free to continue radio-echo sounding in the Otter. We had solved the various teething troubles of the system and were getting consistently good results. Our flight tracks fanned out radially from Adelaide to cover as much ground as possible.

One day we were caught out by darkness. Whereas during much of the summer the sun never sets, in February there can be dark nights. Returning home some hours after sunset from a long flight, we knew that those on the ground would be concerned. As we approached, with only the aircraft's landing light to pierce the gloom, a flickering flare-path suddenly outlined the runway ahead. Fearing for our safety, half the base had turned out to mark the airstrip with burning pots of paraffin.

It brought tears to my eyes. Far from home, and alone in the skies, we often had reason to feel vulnerable. Here was a reminder that for every minute of every hour that we were flying, there was a radio operator and a meteorologist constantly watching over us. Many times during every flight, the aircraft engineers and George Green would check on our progress. Nobody on duty at the radio could go for a meal or even answer a call of nature without ensuring that someone else took over the watch. Aircraft emergencies can happen fast – we might need to get a message out within seconds.

The eleventh of February was a day to remember. The weather was clear to the south as we headed for Fossil Bluff, measuring ice depths along the way. There we refuelled and carried on south down the middle of George VI Sound. Reaching Eklund Islands, we were at the limit of exploration. Turning for home, we laid course over Bach Ice Shelf and Beethoven Peninsula. To our right lay Staccato Peaks. A host of musical names dot the landscape hereabouts – an international symphony of composers and their works commemorated in this remote corner of the world.

The grouping of names was the brainchild of Brian Roberts, who manned the 'Antarctic desk' at the Foreign Office from 1944 to 1975. He saw in group naming an opportunity to soft-pedal the nationalism that had characterized so much of Antarctic naming in the past. What better than to celebrate – without any national bias – contributions to civilization in a variety of fields? Music was one. Other areas of the Antarctic Peninsula commemorate pioneers of aviation, photography, medicine, transport, navigation, glaciology and many other subjects.

As we flew on north, we could see Chopin Hill, Debussy Heights, Dvořák Ice Rise, Franck Nunatak, Gluck Peak, Mount Grieg, Handel Ice Piedmont, Mount Liszt, Mendelssohn Inlet, Mozart Ice Piedmont, Schubert Inlet, Mount Strauss and Mount Tchaikovsky.

Flying low, we came across many features never before recorded. On Wilkins Ice Shelf, there was a phalanx of small ice rises on both sides of our track. I longed to explore the area on the ground.

By the time we had plodded our way home across Marguerite Bay – flying the Otter always felt like plodding – we had been in the air for just over eight hours. We were exhilarated and excited – today we had felt like real explorers, and the radio-echo sounder had performed flawlessly.

Years later I took steps to celebrate what we had achieved by recommending place-names that would put my colleagues permanently on the map of Antarctica. Today, on Wilkins Ice Shelf for all to see, there are Petrie Ice Rises, Burgess Ice Rise and Vere Ice Rise.

At this stage in the season we ran completely out of avgas, having used the total amount that management had allowed for the radio-echo programme. However, John Ayers, still without avtur for his aircraft, had taken to feeding it ordinary motor fuel. This too was in short supply, but he was not to be defeated. A few gallons generally remain at the bottom of every barrel. While most people leave it there because it may be contaminated by rust and water, John had other ideas. Upending each barrel, he poured the dregs through chamois leather into another barrel. When full, this was dragged up to the airstrip and fed to the unsuspecting Porter.

The manufacturers of the aircraft would have been horrified – not to mention our bosses back in London. My only prayer was that the engine would not quit over the sea. Burning fuel for which it was not designed, on every take-off John had to watch the turbine temperature rather than the torque meter. Avgas burns at a higher temperature than avtur. Overheating a turbine engine could damage it in seconds.

Petrie now transferred the radio-echo sounder to the petrol-burning Porter and, on 14 February, we took off for a trial flight. We shared the tiny cabin with a long-range fuel tank that John had made from a used barrel. Superannuated petrol and air bubbles snaked their way over our shoulders in transparent polythene tubes towards the engine. In order to fit everything into the cabin with the mandatory placebo – our rubber dinghy – John had reversed the co-pilot's seat. So I sat facing backwards with a perennially stiff neck from trying to see where we were going instead of where we had been.

Shortly after take-off I smelt burning. John said: 'Too bad, I can't land until we have used some fuel – I am well over maximum landing weight.' David and I were too scared to enquire what would happen if flames appeared. It was the radio-echo sounder that was at fault, so in due course we landed for repairs. My diary entry for the day ended:

> Beautiful clear sunset, all serene, still, peaceful, unearthly, wonderful. Saw the old Southern Cross for the first time at Adelaide.

The next test flight was no more successful than the first. Five minutes after take-off, as we were climbing through 1,500 metres, I saw petrol dripping on to David's shoulder from one of the polythene tubes. I reported this to John, who at once opened the front fresh-air vents while I lunged for the rear window. David tried to tighten a clip on the leaky joint but it still dripped. At this stage a second joint was found to be leaking. John executed a U-turn and power-dived back to Adelaide, where we landed. This time there was no mention of the aircraft being 'above max landing weight'. It was an emergency. John reported feeling the first signs of 'gasoline narcosis'. I threw open the door and breathed deeply.

After checking all fuel joints, John decided on a two-hour wait 'to clear our heads'. Then we were off again. This time there were no leaks, and we completed some excellent sounding runs over the plateau and Wordie Ice Shelf.

Returning to Adelaide after four hours in the air, we observed *John Biscoe* steaming towards the station. Like many young pilots, John Ayers had in him a streak of devilment, and now it came to life. The Porter was equipped with a hatch in the floor of the cabin. It was designed for a survey camera, but now David was instructed to 'open bomb doors!' John lined up on the ship, and at a hand signal, David propelled two rolls of toilet paper through the hole. Both made direct hits – as we might have expected from a military pilot. We later learned that one of them had failed to unroll in the air as intended. It had thudded into the deck beside the Captain. We were informed that he was 'displeased'.

Except for one three-hour flight, the next few days were spent unloading the ship with the year's supplies. Everyone turned to. The most humbling task was carrying sacks of coal from the little jetty to a storage dump. Coal dust permeated lungs and clothes, it stuck to perspiring faces and got in our eyes, making us look like coal miners. At the end of the day we dived into the cold sea hoping that some of it would wash off. Thirty seconds in sub-zero water was as much as I could stand.

John Biscoe sailed on 21 February and the following day we flew for three hours. The Porter was now for the first time able to use its proper fuel, brought in by the ship. But now too, the season was coming to an end. The plan was for John Ayers to stay at Adelaide to bring geologists back from the field with the Porter, while we transferred our equipment once more to the Otter to head north.

Our final magnificent flight was on 27 February up the backbone of the peninsula to Palmer station. Bob Vere was pilot, and Ray Perren, one of the engineers, came along for the ride. Off Palmer we found the US Coast Guard icebreaker *Westwind*, sister ship of USS *Atka* that had carried me from McMurdo to New Zealand seven years before.

Once on the ground at Palmer, I could begin to consider what we had achieved in the course of 75 hours of flying. We had made continuous depth soundings over all types of glaciers found in the Antarctic Peninsula area, including complete cross-sections of Larsen, Wordie and George VI ice shelves both along and across the direction of flow. Ice depths reached 600 metres and were comparable with those of the largest Antarctic ice shelves.[3] I felt well satisfied with the season's work.

We spent two delightful days as guests of Palmer station, being given a guided tour of *Westwind* by Captain Frederick A. Goettel, followed by a half-hour ride in one of his helicopters. Our revelry was cut short by a message from the second BAS ship, RRS *Shackleton*, that they would pick us up later that day. Our equipment was still in the Otter parked at the airstrip 2 kilometres up the slope of the ice piedmont. Much of that was badly crevassed but our American colleagues had set out a line of stakes to mark a safe route. However, there were bridged crevasses within metres of the 'safe' route. David Petrie borrowed a motor sledge for the day. In whiteout, we were driving up towards the aircraft when we lost the trail. Knowing that wandering in any direction could be fatal, I belayed David to the motor sledge. Crawling on hands and knees, he found the trail only because of the slight shadow cast by his bent-over body. Moments like these told on our nerves.

On board *Shackleton*, the older of the two BAS supply ships, we were now homeward bound and in the safe hands of Captain David Turnball, known by all as Frosty but always addressed as Captain. We had an uneventful voyage to Punta Arenas, Chile, whence I flew to Washington, DC.

I was pleased that we had been so successful with the radio-echo sounder and wanted to show the results to Dr Albert P. (Bert) Crary, Chief Scientist of the Office of Antarctic Programs of the National Science Foundation (NSF). Bert was an old friend and a glaciologist like me. He was excited by seeing our cross-sections of the Antarctic ice sheet and said that he had discussed with Gordon Robin the possibility of mounting an SPRI radio-echo sounder in a US aircraft the very next season.

This was the start of collaboration between SPRI and NSF on radio-echo sounding that was to last for more than a decade. Over 1,000 hours of flying in four-engined aircraft were devoted to studying the depth of ice over vast areas of the Antarctic ice sheet, leading to the production of maps of ice thickness, surface elevation and bottom relief.[4]

That field season might have been my last. Both aircraft reached Deception Island in March 1967, where they were to be overhauled during the winter. The Otter had not had a Certificate of Airworthiness for some years because nobody in the Antarctic was authorized to issue one. On careful inspection, the engineers found major cracks and a probable fatigue failure of the tailplane. John Ayers, by then Head of the Air Unit, cabled the Director as follows:

> Have grounded Otter . . . as no longer airworthy . . . By the grace
> of God we have been fortunate not losing aircraft through fatigue
> failure in the air. One more flight might have been enough . . . [5]

Colleagues report that my face turned white on taking in the message.

4. NORTHWEST PASSAGE (1969)

The first ever commercial cargo

It has long been known that there are oilfields on the North Slope of Alaska, but there was no commercial interest because oil was easier to come by elsewhere. The picture changed in the 1960s when the Suez Canal was closed and the US Government sought to reduce oil imports. Exploratory drilling revealed a bonanza awaiting anyone bold enough to face the high costs of working in Alaska's extreme Arctic environment.

British Petroleum (BP) had the courage and the vision. After a massive investment the company succeeded in bringing up oil in the Prudhoe Bay area in 1968. The first well flowed at the rate of 2,400 barrels a day and it became clear that there were very substantial reservoirs of exploitable oil. Some estimates ran as high as 25 billion barrels. The State of Alaska responded by leasing large tracts of state land, which brought them a tidy windfall of around a billion dollars. Later they were to regret that they had leased the land so cheaply.

A more intractable problem was how to get the oil to market. In February 1969, a consortium of oil companies proposed laying a pipeline all the way from Prudhoe Bay to the south coast of Alaska, although it would have to overcome two mountain ranges, an active fault belt, rivers, swamps, permafrost and – worse still – discontinuous permafrost. The tundra, frozen in winter, turns into a quagmire of shallow lakes and streams in the short summer season. In spite of the difficulties, the consortium applied for a permit in June 1969 and, anticipating it, ordered from Japan 1,300 kilometres of 1.22-metre-diameter steel pipe to carry oil to the ice-free port of Valdez.

However, it was believed that a permit to build the pipeline might not be given unless alternative routes to market had been considered. The second option was to take the oil by sea, not west through Bering Strait but east through the Northwest Passage to the Atlantic, where there was a better market. The problem was pack ice – in places heavier than on any existing trade route. No ship had ever carried a commercial cargo between the Pacific and the Atlantic by this route.

The largest ship that had ever ventured into pack ice was the Soviet nuclear-powered icebreaker *Lenin* with a displacement of 16,000 tonnes. It would not be safe to extrapolate from her performance to that of a modern tanker displacing ten times as much. An experimental voyage with a big tanker was needed to establish design criteria. Market pressures made the need urgent. However, projected costs were such that only a consortium of oil companies could stand the risk.

Humble Oil and Refining Company, a subsidiary of Standard Oil of New Jersey (Esso – now known as Exxon), chartered the tanker SS *Manhattan* from her owners, Seatrain Lines. They sought $10 million to convert the ship for ice-breaking. Offering equal access to the results for a fee of $2 million, they reckoned to bring in other oil companies, among them BP, Atlantic Richfield, Mobil, Phillips Petroleum, Union Oil, Gulf Oil, Getty Oil, Amerada Hess and Louisiana Land. In the event, only BP and Atlantic Richfield took the bait, cannily negotiating fixed-price contracts. Esso lived to regret their gamble when the actual costs escalated to $40 million.

I came into the picture in December 1968 when the Scott Polar Research Institute was asked to advise on the types of ice likely to be encountered. In return for its $2 million contribution to Esso, BP was to be allowed three observers on board. Their principal representative was Captain Ralph Maybourn, Chief Marine Superintendent of the BP Tanker Company. Paul Heywood, a naval architect, was second. Having covered those essentials, they wanted an ice man, so I volunteered. I had stayed in some expensive places but the idea of a shared cabin for $2 million was something new.

BP was a good company to work with and insisted on thorough preparations. In June 1969, while *Manhattan* was being converted, I was sent to Toronto for consultations with the Canadian Department of Transport; they were to be responsible for observing Arctic ice conditions throughout the summer from aircraft. Next I went to the US Army Cold Regions Research and Engineering Laboratory in Hanover, New Hampshire, to learn about their plans for experiments on the physics of ice; and finally to Houston, Texas, for discussions with the overall Project Manager Stanley Haas and his Technical Coordinator Abraham Mookhoek. Neither of these gentlemen had any experience in pack ice, but I was impressed by their grasp of the very wide range of factors affecting Arctic operations.

Competition between the collaborating companies reared its head when Haas explained that I must consider the bridge, engine room, instrument and analysis rooms, and even the ice laboratory, out of bounds. As a glaciologist of 20 years' standing, this was a rude slap in the face. I was to encounter further security problems, but put them down to sour grapes because BP had secured their nominally equal partnership so cheaply. However, on the voyage it proved impossible to police the restrictions, so I saw everything that I wanted to see.

SS *Manhattan* offered a virtual catalogue of maritime superlatives. Launched in 1961, she was at the time the largest commercial vessel flying the United States flag. Weighing 31,000 tonnes empty, she could be loaded to displace more than 150,000 tonnes. Her immense size meant that, laden, she was too deep to enter New York harbour and many other ports. In the course of her life she carried not only crude oil but millions of tonnes of grain to India, Pakistan and the USSR, becoming the largest ship to pass through the Dardanelles to the Black Sea. I often wondered how the bread made in these countries could avoid a faint tang of crude oil.

In order to meet the deadline for an Arctic tanker test in 1969, the ship was cut into four pieces in Chester, Pennsylvania. While the stern section remained at Chester for ice strengthening, the midships section was towed to Mobile, Alabama, and the forward section to Newport News, Virginia. An entirely new 75-metre-long ice-breaking bow was built at Bath, Maine. In all, thousands of dockyard workers swarmed over the various pieces 24 hours a day. The ice strengthening consisted of a 2.5-metre wide reinforced 'ice belt' on each side of the hull which increased the overall width at the waterline to 45 metres. The bow section had a width of 47 metres, the added 2 metres being an attempt to force ice clear of the main part of the hull. There was additional internal strengthening against ice pressures. A helicopter deck was built and the number of cabin berths was doubled. The two conventional propellers were replaced with special bronze-nickel propellers 8 metres in diameter.

When the five parts of the ship were finally assembled, her overall length had grown from 287 metres to 306 metres. Altogether some 9,000 tonnes of steel were added in the conversion.

There was one thing that could not be changed – the steam turbine engines. Theoretically developing 43,000 horsepower, they were the pride of the tanker fleet but much underpowered for ice-breaking. About one-tenth of their boiler output of steam was needed for auxiliary and heating systems, leaving only 39,000 horsepower delivered to the shafts. However, by measuring performance within these limits, it should be possible to extrapolate to the power that would be needed for year-round operation of a vessel twice the size. Tankers displacing 500,000 tonnes were being built at the time.

The phenomenal transformation of *Manhattan* was big news in the US but it was not the greatest event hitting the headlines. Neil Armstrong stepped from *Apollo-11* onto the surface of the moon while the five new or rebuilt hull sections were being welded together. The reborn *Manhattan* set sail from Philadelphia on 23 August 1969. If ships could stand on end, she could now almost top the Empire State building. Of the more than 100 persons on board, only 54 would be needed to operate the ship as a tanker. Esso staff took up the lion's share of the accommodation. Ralph Maybourn and Paul Heywood were there for the westbound voyage but I would join later for the eastbound passage from Point Barrow, Alaska – assuming the ship ever got there. The eastbound voyage in late September might be more difficult in that newly forming ice could lock the older ice floes together.

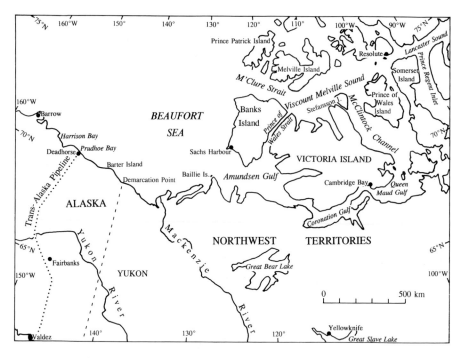

Map 4   The Northwest Passage (western section)

Conventional icebreakers, described by one participant as 'rather diminutive watchdogs', stood by in Davis Strait to accompany the ship and to help if needed. One of them was the Canadian CCGS *John A. Macdonald*. The other was the American USCGC *Northwind*, which had come from the Pacific earlier in the summer and, accompanying *Manhattan* westbound, became the first ship ever to transit both ways in one season. With luck, *Manhattan* would become the second.

I followed *Manhattan*'s progress in the newspapers. So great was world interest in the voyage that 100 people, mostly newsmen, chartered a jet in Montreal and flew more than 6,000 kilometres just to circle the ship for a few minutes while she was moving through ice.

There were no real problems before reaching heavy old ice in the western half of Viscount Melville Sound. The biggest difficulties came when she tried to sail through M'Clure Strait into the Beaufort Sea – where no deep-draft ship had ever been before. This route was attractive as the shortest passage to Alaska, but multi-year-old ice drifting east from the Arctic Ocean had created an ice jam between Banks Island and Melville Island. Turning, she steamed down Prince of Wales Strait into Amundsen Gulf and had no serious difficulty after that.

I flew to Anchorage, Alaska, across the Arctic Ocean on 19 September.

As luck would have it, I was travelling first class with Sir Frederick Harmer, a BP director, and Commander Edward Platt, the Technical Head of BP Tankers. After a night in the Captain Cook Hotel, we flew north across Alaska, enjoying a grand view of Mount McKinley – the highest peak in North America (6,194 metres) – to our left. We were in a six-seat Merlin turboprop aircraft with a single pilot, so I asked to sit beside him. As we passed over Fairbanks, a message crackled into the headphones from Air Traffic Control 7,000 metres below: 'Please tell Dr Swithinbank that Carl Benson is disappointed not to have a visit.' Carl was an old friend – but I had thought we were travelling incognito. I now concluded that everyone in Alaska must be following the *Manhattan* saga.

We landed at Deadhorse airstrip just inland from Prudhoe Bay. Not long before, there had been nothing in this place but virgin tundra dotted with small lakes. Now there was a runway consisting of gravel 1–2 metres in depth excavated from stream-beds and laid straight on top of the marshy tundra. There were serried ranks of accommodation units joined together, each unit designed to fit in Lockheed Hercules aircraft because there was no road connection with Southern Alaska. We could see a number of drilling rigs in the distance.

After a night's sleep we went to one of BP's massive rigs, where mud-spattered roustabouts were busy lowering and connecting drill rods. Everyone was tight-lipped about whether or not this particular well had struck oil. Then we were driven to a gravel dock bulldozed into a very shallow part of the Beaufort Sea. This was for barges bringing oilfield equipment from the port of Seattle. The nearest water deep enough to float *Manhattan* was 16 kilometres offshore. Nobody had solved the problem of how to load deep-draft tankers held fast in drifting pack ice.

The following day we flew 300 kilometres west along the coast to Barrow, the northernmost settlement in the United States. It is an old Eskimo village with a long tradition of whaling but now transformed by a sudden influx of outsiders offered inflated wages in oilfield-related industries. That day there was also a flock of journalists covering the arrival of *Manhattan*.

Our ship was visible as a sleek black silhouette some way out to sea. A helicopter came for us, and Ralph Maybourn was on the flight deck to meet us. The newcomers were ushered into vast cabins within the midships superstructure. Mine was said to have been outfitted for the shipowner's mistress. I shared it with Bern Keating, a 'hack writer', to use his own words, commissioned to produce a book about the voyage, and Dan Guravich, a freelance photographer who would illustrate the book.[1]

The ship was riding low in the water, partly because she was deeply ballasted and partly because her fuel order of 184,000 barrels of bunker oil at the start of the voyage was 'the largest ever placed for a commercial vessel'.[2]

Unusually for a ship at sea, there were three captains on the bridge, with an additional pair of 'guest' captains. The company captains were Roger Steward – the Master – assisted by Don Graham and Arthur Smith. All three were experienced tanker skippers and had observed Arctic operations from icebreakers. One of the guests was Captain Thomas C. Pullen of the Royal Canadian Navy, the very same Tom Pullen who commanded HMCS *Labrador* when I was on board in 1956. The other was Captain Frederick A. (Beef) Goettel of the US Coast Guard. I had known Beef since we met in the icebreaker USCGC *Westwind* at Palmer station in February 1967. Each guest captain was serving – or so he hoped – not only as official observer for his government but also as 'Ice Master' to advise on the technique of ice-breaking. In the event, I noted that the company captains took very little notice of the advice of the guest captains.

The ship had left Philadelphia with supplies for six months: 32 tonnes of canned and dried food, 23 tonnes of meat (including 764 turkeys, 789 ducks and 424 chickens), 18 tonnes of fruit and vegetables, 3.6 tonnes of potatoes, 2.7 tonnes of coffee, 7.5 tonnes of milk, 300 watermelons, 90 kilos of peanut butter, 1 tonne of ice cream and 51,000 eggs. The officers, crew and passengers were fed in a giant dining room situated aft over the propellers. At mealtimes we certainly knew when the ship was breaking ice. Chattering of the two giant propeller blades as they sliced into ice floes could spill drinks and shake plates off the table. We held on tight because the food was too good to lose.

The next morning, 23 September, saw us steaming eastwards past Harrison Bay, Prudhoe Bay, Barter Island and Demarcation Point – the border post between Alaska and Canada. I had known these names from the time – 13 years before – when I was studying the distribution of pack ice in the Northwest Passage. Indeed, it was on the basis of that experience that I had been asked to join *Manhattan*.

In the course of the earlier study I had read dozens of logbooks of trading vessels that had fought their way along the uncharted shallows between the pack ice and the shore. Now, instead of needing 2 metres of water to get by, *Manhattan* needed more than 16 metres to avoid any risk of grounding. Even in 1969, the charting of these waters was inadequate, and on one occasion the echo-sounder indicated that a seabed pinnacle rose suddenly to depths uncomfortably close to the keel. The captain on watch was visibly shaken, because with the momentum of our passage there was no chance of stopping in a reasonable distance by ringing 'full astern' on the engine telegraphs.

The icebreaker CCGS *John A Macdonald* was accompanying us on behalf of the Canadian Government – to watch our progress and to help as necessary. Icebreakers can be described as very heavily reinforced ships with engines yielding more than about one horsepower per tonne of displacement. 'Johnny Mac', as she was known by her crew, displaced 9,000 tonnes; her diesel-electric plant could develop 15,000 shaft horsepower. Captain

Porter take-off at Adelaide station

Bourgeois Fjord seen from the Pilatus Porter

Cramped quarters at Fossil Bluff

Dinner at Palmer station on Norsel Point, Anvers Island

SS *Manhattan* in Viscount Melville Sound

Time to jump out of the way. *Manhattan* was 47 metres wide

The forecastle could almost accommodate a football field

There was space for another football field behind the bridge. Two icebreakers followed in our track

Paul Fournier invited Edward Platt, Tom Pullen, Ralph Maybourn and me to lunch. Using our helicopter, it took just one minute to fly from one ship to the other. Fournier was a crusty French-Canadian sea dog from the Gaspé Peninsula who knew the Arctic like the back of his hand. After a fine meal and a tour of the ship we flew back in *Macdonald*'s helicopter.

Our second escort, USCGC *Staten Island*, was 82 metres long, displaced 7,000 tonnes, and was powered by diesel-electric machinery like *Macdonald*. Smaller than *Macdonald*, she was one of seven 'Wind' class icebreakers constructed between 1942 and 1946. Four were built during World War II and named after the cardinal points of the compass. When more icebreakers were needed, the authorities faced a naming dilemma. Rather than launching a Northeastwind, and so on, they resorted to other names. Like some fickle woman, *Staten Island* had changed her name three times. She began life in 1944 as the Coast Guard cutter *Northwind*. Immediately on commissioning she was loaned to the Soviet Union and renamed *Severniy Veter* (North Wind). Returned to the US in 1951, she became USS *Northwind*. In 1952 she was renamed USS *Staten Island* to avoid confusion with the second *Northwind* – of later vintage – the ice-breaker which had accompanied *Manhattan* on her westbound voyage. Finally in 1966 she was returned to the US Coast Guard.[3] *Staten Island* had 12,000 shaft horsepower – nearly 2 horsepower per tonne – but in heavy ice still had to fight every inch of the way.

Manhattan was breaking an impressive channel through a 13-kilometre-wide consolidated multi-year ice floe but after some time was brought to a standstill. Tom Pullen called *Macdonald* on the radio, asking: 'Would you mind coming over to nibble about our quarters?' Backing and ramming, the small icebreaker slowly worked herself alongside *Manhattan* to release the pressure. Enquiring why *Manhattan* seemed unable to move astern without assistance, I was told that she could only apply one-third of her power to the propellers when going astern. Evidently this is a normal and acceptable design limitation in steam-turbine-powered ships because the only time they need to go astern is in emergency or when docking. Real icebreakers, on the other hand, are designed to have all their engine power available ahead or astern. Full astern power would certainly have to be designed into any operational ice-breaking tanker.

Manhattan's lack of astern power caused no end of misunderstanding with representatives of the press, and afterwards with scientists and others whose opinions were based on press reports. Seeing us being assisted, they concluded that tankers could never proceed through pack ice without help from icebreakers. The truth is that while steaming ahead, *Manhattan* outclassed both of her accompanying icebreakers. Neither could move through vast unbroken floes without continually backing and ramming, achieving perhaps 100 metres with each forward thrust. Sometimes they had difficulty even following us because the jumble of floes *Manhattan* left in her wake would block the channel. At no time on the voyage were we *led* through ice.

The next time we approached difficult ice, the ship stopped and a helicopter was launched to search ahead for leads. A Canadian Government DC-4 flew over us and reported on ice further ahead. Edward Platt and a number of journalists left the ship on 25 September at Sachs Harbour on Banks Island. Some Eskimos came aboard to sell sealskin clothing. Earlier, on the westbound voyage, the local school had been specially closed for the day and newsmen had passed the hat to charter a DC-3 so that Eskimo children could fly over the ship.

Now we entered Prince of Wales Strait, a 270-kilometre-long channel separating Banks Island from its neighbour to the east – Victoria Island. At first passing easily through first-year ice 1–2 metres thick, we later had to battle close-packed consolidated floes. I was delighted to have a visit from Moira Dunbar, the same Moira Dunbar who had overseen my work at the Defence Research Board of Canada 13 years earlier. She was spending a few days on board *John A Macdonald* as a guest of the Captain. For the first time we had to stop for the night because of darkness. Autumn was now truly on its way. A film, *The Ugly American*, was laid on for us, after which everyone had the first really peaceful night's sleep since we left Barrow.

As we emerged from the strait into Viscount Melville Sound, the sun shone on a brilliant scene of dense pack ice peppered with new snow. Gone was the patchwork of melt-puddles and dirt-sprinkled ice through which we had been steaming. Here we would search for the severe ice conditions needed for the most crucial experiments. Icebreakers proceed not just by crushing ice but also by forcing floes apart. But that would not provide the data needed for our purpose. We must find a single vast floe – the nearest equivalent to the unbroken expanses of landfast ice that future ice-breaking tankers would have to steam through in winter. It must be too large to split – we must break a path.

It was not hard to find. In fact the DC-4 had reported by radio that there were several floes 1–5 kilometres across in the vicinity. Now there developed a routine that was repeated every day for the next couple of weeks. The ship nudged into the icefield – usually 3–6 kilometres across – and stopped. A ladder was put down to the ice. Groups of glaciologists – myself included – went down it with ice drills, bamboo stakes topped by flags, and a plethora of instruments to test the mechanical, chemical and thermal composition of the ice. Each group carried a hunting rifle in case bears challenged our right to invade their domain. We set out the bamboos in a long straight line for 2 kilometres ahead of the ship. This was the track that the Captain would follow – after we had finished.

Every 30 metres along the line, we drilled with a hand auger to determine the thickness of the ice. Most of the old floes were 1–2 metres thick but some pressure ridges went so far down that we never did get through them.

The drills brought up cylindrical samples that were used to measure the compressive strength of the ice, its temperature and brine content. This time, and on many other occasions, the whole procedure took six hours in a biting wind. Standing on the ice ahead of the ship, we marvelled at how, with *Manhattan*'s stupendous beam, she could make any progress at all. Weighed down with sea water ballast, she was riding so low that it made her width seem even greater. On the walk back to the ship we saw polar bear and fox tracks. I rehearsed how to confront the bear with an ice auger but fortunately acquired no 'experimental data'.

The final act was to recover everyone from the ice, back off a little and then charge the line at full power. *Manhattan* shuddered with the impact of each ridge and hummock. Her 306-metre length was so great that there was a visible flexing of the hull as the raked bow tried to rise up on top of the ice. But the tremendous mass of the ship meant that the ice was forced down under the bow instead. Plumes of salt spray shot 20 metres into the air and drenched anyone standing on the forecastle. Sighting over the bow from the midships superstructure, I could see that even with the heaviest impacts, the bow rose no more than half a metre. Flexural waves could be felt racing aft through the hull; I timed them at a rate of 1.8 cycles per second. The sensation was like earthquake waves superimposed on the ice shocks. I began to wonder whether the ship was designed for such abuse. Tankers, after all, have been known to break in half.

Below decks, the laboratories were buzzing with activity. Four hundred strain gauges were spread throughout the ship's structure to measure its 'elastic' deformation under the impact stresses. At least we hoped for elasticity rather than any permanent bending of the hull. Hundreds of thousands of dollars has been spent on elaborate recording devices to measure everything that might be relevant to the design of future ice-breaking tankers of 300,000 tonnes or more. One thing the instruments could not do was measure our speed through the water. Any external fitting would have been wiped off the hull by ice in short order. We could – and did – measure speed over the seabed by recording the Doppler shift of sonar pulses transmitted ahead, astern and to either side.

While charging through the line of bamboos, the forecastle was awash with staff armed with stopwatches and notebooks, recording exactly where we were along the line so that the drag on the hull could afterwards be analysed in relation to the measured ice thickness and its physical properties at that point. A delightful anachronism was the sight of learned scientists noting the speed of the ship by the very same means that mariners in the days of sail had measured it. Men stood at each end of a 40-metre-long fore-and-aft line on the forecastle. The front man threw a piece of wood – any piece – onto the ice opposite where he was standing. As he did so, the aft man clicked a stopwatch, waited until the wood was abeam of where he was standing, then stopped the watch. Simple arithmetic gave him the speed of the ship through the ice.

To add mass in order to simulate the response to ice of an even larger ship, the cargo (oil) tanks had been filled with sea water right up to deck level, in fact to the rim of the small hatch above each tank. We could see water spilling over. This meant that the loaded displacement of the ship was a whopping 155,000 tonnes, about twice the mass of the largest liner afloat.

A Humble press release waxed lyrical about 'The most talked about ship in the world today'. Tom Pullen commented: 'The Manhattan has put on the most awesome display of icebreaking I have ever seen.'[4]

5.   HOW TO BREAK ICE (1969 continued)

With the largest ship ever to try it

On 30 September we were joined by Canada's newest icebreaker CCGS *Louis S St-Laurent*. She was on her maiden voyage and had come to keep an eye on us and – we suspected – to show off her ice-breaking prowess. Now we had three icebreakers, normally following in our wake but always available – and often used – to chop us free when we were unable to move astern. *St-Laurent* out-performed the others. She was at the time the most powerful operational icebreaker in the world, so that was to be expected. *Lenin*, the Russian nuclear-powered icebreaker, had suffered a reactor meltdown and was temporarily out of action.

Looking over *Manhattan*'s bow during one of her longer thrusts through a floe several kilometres wide, I was surprised to see that the stem was not itself breaking ice. Our bow wave was doing the breaking some 10–20 metres ahead of the ship by arching the ice upwards and causing multiple fractures. The greater the speed, the greater the convexity of the standing wave and the greater its contribution to ice-breaking. The stem, which sloped at the unusually gentle angle of 18 degrees – then pushed the debris beneath the ship. Debris may be the wrong word because many an ice floe 10 metres across slid quietly underneath the hull and travelled the entire length of the ship before being mercilessly chewed by the propellers. Other floes came out astern intact but with giant saw-tooth gashes where each blade had left its mark.

What happened right in front of us was only part of the story. Aft of the newly built bow, *Manhattan*'s hull was shaped like a bathtub. But this was a 47-metre-wide bathtub, meaning that a great deal of ice was being pushed sideways against unbroken ice. Ice being only slightly lighter than water, it is comparatively easy to push down beneath the hull. Pushing it sideways is costly in energy terms. The channel astern of the ship was not just 47 metres but as much as 60 metres across.

Chief Engineer Colos Bennett was concerned about the new bronze-nickel propellers because they were softer than steel. Our homecoming depended on their survival under the constant onslaught of ice. We had thousands of times felt the ship shudder as the blades cut into ice floes. *Staten Island* kept qualified divers on board, so we invited them to go down to make an inspection. *Manhattan* was stopped in a giant ice floe, but the stern was surrounded by debris through which an opening could be made by nudging pieces apart with oars from the divers' inflatable boats. It was an unenviable task to drop into water at –2°C. We saw their faces contorted with pain as they submerged. A line was attached to each diver's waist belt for safety. They surfaced after about ten minutes to report that all was well except that the thin tip of each blade was bent over and rolled up against itself. The engineers were greatly relieved. Hardened steel propeller blades have been known to snap off from the impact of hard ice. Deformed tips were a small price to pay for the survival of ours.

Ralph Maybourn was interested in looking inside the bow structure during one of our ice-breaking charges and invited me to come with him. Entering through a hatch on the forecastle, we descended a series of ladders to the level at which ice floes were striking the hull. The noise was deafening as each impact reverberated in the enclosed space. Neither of us had ever experienced an artillery barrage but we concluded that it must sound much the same. As a naval officer in 1945, I had been in charge of a gun turret in a cruiser. From memory, I felt that intermittent explosions of cordite were less of an assault on our ears than the barrage we were now experiencing.

Construction of the bow had been so hurried that it was unpainted. The high-tensile steel plating was 40 millimetres thick. However, it was the massive frames supporting it that I could not have imagined. Not only were they very closely spaced but each one extended a metre or more inboard from the skin. Extra bracing supported each frame. It seemed as if some shipyard welder had stood back and thought, 'Ah, I can get rid of an odd scrap of steel by tacking it across this gap!' There was no doubt that the naval architects had done everything possible to prevent even the hardest of ice floes from penetrating the forepeak. To the best of my knowledge they were successful, the structure even surviving one occasion when the Master – unwisely I thought – tangled with an iceberg.[1]

Some of the floes with big pressure ridges caused real trouble. We made only 30–50 metres with each forward thrust. But already some of the staff were beginning to scale up, in their minds, to the kind of power they would need for year-round operations. The Technical Coordinator, Abraham Mookhoek, gave a lecture which clearly showed his optimism. He said it would cost $125 million to build a shipyard capable of turning out two to four 300,000-tonne ice-breaking supertankers per annum. Each ship would cost $60 million and Esso wanted first deliveries in early 1973.

On 6 October, there was a classic demonstration of a problem inherent in the use of long, parallel-sided ships in ice. We were heading due south in consolidated and hummocked multi-year pack ice. In attempting – and failing – to turn north, we were forging into uncharted waters and still heavier ice. The ice alongside simply prevented the stern from moving sideways in response to the rudder. All attempts to back up in the channel and set off on a different track failed. Finally, *Macdonald* came up on the port side 100 metres away and chiselled at the ice while *St-Laurent* came up on the starboard side to do the same. The newer and more powerful ship gave a superb demonstration of brute-force ice-breaking. But with two ships at such close quarters, Captain Steward was terrified of collisions. Ice, however, serves as an effective fender. Only by breaking a ring around us was it finally possible to turn north, albeit in several stages.

Air temperatures had been generally mild – around 0°C – but now with the onset of autumn some days brought –10°C or colder. The hand-drilling and experimental work on the ice was still pleasant except when the wind cut into our faces and made fingers and toes go numb. Vigorous exercise was the recipe for survival until we were back on board.

The ship stopped for the night in a large floe within sight of Melville Island, but when we tried to move in the morning nothing happened – we were stuck. *Manhattan* was fitted with a very powerful heeling system using 2-metre-diameter pipes across the hull. Two thousand tonnes of water could be pumped across to heel the ship from 3 degrees on one side to 3 degrees on the other in the space of 70 seconds. We used the system but still needed help from the icebreakers.

I learned that there were altogether 500 people in the fleet: 127 in *Manhattan*, 96 in *John A Macdonald*, 92 in *Louis S St-Laurent* and 185 in *Staten Island*. The contrast between the crew sizes in the last two was partly explained by their vintage – the American ship being 25 years older than the Canadian. This probably reflected the change in crew sizes made possible by design improvements dictated by rising labour costs.

Day after day we repeated the experiment of measuring the physical properties of vast ice floes before charging through them. However, my diary noted:

> The testing programme is very poorly organized and many opportunities are wasted.

One of the helicopters had been written off on the westbound voyage because its inexperienced pilot landed on a newly frozen melt pool and dropped through into the sea beneath. Nobody was hurt and the second helicopter continued flying. Sometimes it circled the ship to film our progress while charging ice.

The uncooperative attitude of the management that had shocked me on my visit to Houston showed no signs of abating. The Project Manager Stan Haas refused my request to occupy a spare seat in the helicopter and the same day refused to allow Paul Heywood to visit *Macdonald*. However, he was less successful in policing what went on below. It is against the instinct of scientists not to share information and I was invited into almost every laboratory that was off-limits. I had a tour of the engine room – another forbidden fruit – with one of the engineers. One palatial compartment held serried ranks of oil-fired boilers, another a number of giant steam turbines. I did not envy the people who had to work down there.

On 9 October we were 30 kilometres north of Stefansson Island, named after Vilhjalmur Stefansson, one of the great explorers of this corner of the Arctic. Stef, as he was known to his friends, had helped me in 1957 when I was collecting historical records of ice distribution in these latitudes. His magnificent polar library in Hanover, New Hampshire, was a mecca for scholars of Arctic affairs. He died in 1962 at the age of 83.

The following day I went to lunch on board *Louis S St-Laurent*. The Captain explained that with a displacement of 13,300 tonnes and 24,000 shaft horsepower his ship could make 17 knots in open water, which is rather fast for an icebreaker. Her steam turbines powered electric propulsion motors that were geared to triple propellers. Air bubblers could be used to reduce the ice-drag on the hull plating. I have always loved the smell of a new ship and it was obvious that the crew were proud of her.

By now I felt that I had a good understanding of *Manhattan*'s performance in ice. Moreover, I had concluded that the season was neither unusually tough nor unusually easy so far as ice was concerned. There seemed little point in staying to witness further experiments repeating the same routines. *Manhattan* was destined to work her way eastwards towards open water in Lancaster Sound, after which there would be no more ice on her voyage that was scheduled to end in New York. Paul Heywood and 16 other executives, staff or observers had come to the same conclusion. Learning that *St-Laurent* was to go to Resolute to pick up spares for the fleet, we packed and transferred to her on 12 October. The Canadians proved to be generous hosts, quite unperturbed by looking after 18 guests.

The icebreaker steamed east through Melville Sound; there was a full blizzard, and some very close pack ice, but nothing presented any real difficulty. Nevertheless, we stopped for the night. *St-Laurent* had powerful searchlights facing ahead, but even with these, we might battle ice that – with distant views – could have been avoided.

It took two days to steam the 400 kilometres to Resolute, the Eskimo village on Cornwallis Island, where there is a Canadian government-operated airstrip and weather station. The settlement serves as the main base for air operations over the Queen Elizabeth Islands, a vast archipelago north of this latitude and stretching from the Beaufort Sea in the west to Baffin Bay in the east. There are only a couple of other weather stations in the whole area and some low-key commercial oil-prospecting operations. The only other human presence occurs in summer, when geologists and other scientists are spread out over the islands to pursue their calling. All of them are supported by small aircraft from Resolute.

However, our arrival here was to make connections south to the mainland rather than north to the wilderness. Resolute is the only airport at this latitude with regular air services to Montreal by way of Iqaluit on Baffin Island. After waiting some days for the scheduled jet to arrive, we thought that 'sub-regular' would be a better description of its timetable. I was happy with the delay because it gave time to see the ice reconnaissance DC-4 that had been flying over us and to talk about ice conditions with experts. I also met Welland (Weldy) Phipps, Canada's foremost Arctic bush pilot. He owned Otter and Super-Cub aircraft capable of landing on unprepared tundra, beaches and even smooth mountain tops. He had developed the use of oversize low-pressure tyres to land on rough and rock-strewn surfaces. Weldy began by using Otter tyres on Super-Cubs and DC-3 tyres on Otters, but had graduated to specially made lightweight tyres of similar dimensions. His exploits were legendary and he offered to fly anyone anywhere – at a price.

Meanwhile the oil company executives who had come ashore from *Manhattan* were bored with the slow pace of life and became more anxious by the hour. Among them was Sid Wire, General Manager of Humble Oil Company's Marine Department. Waiting for anything was not his style. After a gloomy weather forecast for the 'scheduled' jet, Phipps' Norwegian pilot and Austrian co-pilot offered to fly a Twin Otter to Yellowknife for $4,000 in cash. When Mr Wire capitulated, he was immediately surrounded by a crowd of aspiring hitch-hikers. I was one. The happy outcome was that 12 of us were allowed aboard with all our baggage.

It was a magnificent 700-kilometre flight across Prince of Wales and Victoria islands to Cambridge Bay. After refuelling there we continued over 850 kilometres to Great Slave Lake. Passing over endless tundra and taiga devoid of settlements or radio aids, I marvelled at the pilots' ability to navigate with confidence. But Yellowknife duly appeared at the appointed time. On landing, we rode a fleet of taxis to the Gold Range Hotel.

Yellowknife, with a population of about 4,000, was the largest town in Canada's Northwest Territories. It was founded in the 1930s as a gold-prospecting and mining centre. A small tourist industry had recently added to the population. We spent the evening eating Arctic char and communing with the locals in a bar.

The next day our party scattered to destinations far and wide. Having experienced the inter-company tensions on the voyage – and even some intra-company tensions, though none within the BP group – we had never been part of a single team, so we parted with warm but perhaps not fond farewells.

Manhattan's 16,000-kilometre voyage ended on 12 November 1969 with a fireboat welcome in New York harbour, 80 days after her departure from Philadelphia. She had to pump overboard all the ballast that had kept her deep in the water for ice-breaking. Later, in dry dock, a hole was found in the ship's side plating 'big enough to drive a truck through'.[2] The Arctic ice had left its mark. No. 4 wing tank had been holed well below the waterline and below the reinforced ice-belt that had been added to the hull. The ice-belt itself was undamaged.[3] The plating perforated by the ice was 25 millimetres thick. The ship's engineers must have been aware of the damage in the course of the voyage but they never let on. I was comforted to realize that as the tank had been in ballast, sea water would have flowed out – not in.

Manhattan was the twelfth ship in history to transit the Northwest Passage. She was 3,000 times the size of the first ship to complete the passage – Roald Amundsen's *Gjøa* in 1903–06.[4] She was also the first to carry a commercial cargo: a single, symbolic, gold-painted barrel of Prudhoe Bay crude oil. The owners were proud of it. Stan Haas was quoted as saying: 'The transportation cost of the golden barrel we have aboard our ship is rather high . . . ' At \$40 million, this was a rare moment of understatement.

Finally, my own conclusion was that tankers capable of all-season transits of the Northwest Passage without icebreakers could be developed by extrapolation from the *Manhattan* experimental results. This view was shared by most of the naval architects, marine engineers, oil-company executives and glaciologists involved.

It was not, however, shared by those whose opinions were formed on the basis of press reports of *Manhattan*'s need for icebreaker assistance. One newspaper asserted:

> The consensus of Ottawa officials is that the Manhattan project, aimed at finding an economic route through the Northwest Passage, has failed.[5]

Humble Oil Company thought otherwise. Some months later they awarded a contract to Newport News Shipbuilding and Dry Dock Company to design a giant ice-breaking tanker. They sent *Manhattan* on a second voyage to Baffin Bay to test her performance under winter ice conditions. The voyage took place in April and May 1970, the period of maximum ice cover. BP invited me to go along but I felt that, at that stage, it was more important to send naval architects to consider the design implications of the tanker's performance.

Manhattan was eventually returned to her owners, Seatrain Lines, in July 1970. Under the terms of the charter, Humble had agreed to remove the ice-breaking bow and other special equipment. In the event – and presumably with an eye to the future – Seatrain decided to preserve the ship's ice-breaking capability.

The Chief Scientist of the US Army's Cold Regions Research and Engineering Laboratory declared: 'The *Manhattan* not only broke ice, she dislodged scepticism in the mind of man concerning the Arctic. Now everyone is interested, and this, ultimately, may be the *Manhattan*'s greatest contribution.'[6]

My report to BP[7] began by noting that it is an axiom of meteorology that weather, in the eyes of the local population, is always exceptional. So it is with ice. Some Alaskans told me that 1969 was an exceptionally good ice year off the north coast because the ice was well clear of the coast in late September. But this situation has occurred in the past and there are records of the ice edge being considerably further north. Throughout the Northwest Passage in 1969 the ice was well within the normal range of variability.

After describing the performance of the ship in ice, I launched boldly into suggestions relating to the design of ice-breaking tankers. I asked what bow shape would push the ice downwards against only water resistance; would avoid upending ice floes; would avoid crushing them against the ship's side; would avoid forcing debris sideways into the unbroken floe; and would avoid the frictional drag of debris confined between the ship's side and unbroken ice.

My suggestion was a gently sloping square-ended bow like a river punt, but faired at the sides. The optimum slope would have to be decided by model tests. A 20-degree stem angle would place the forefoot 55 metres aft of the waterline at the bow. This would involve some loss of seaworthiness and an increased risk of pounding in heavy seas.

Naval architects politely advised me to confine my comments to subjects for which I was qualified. However, years later, I had some satisfaction when a bulk carrier displacing 38,000 tonnes (28,000 tonnes cargo capacity), MV *Arctic*, was launched with essentially the hull form that I had described. In 1986 she was given a new bow to improve her ice-breaking performance. Its stem angle was 20 degrees.[8]

I also proposed small twin bilge-keels to reduce the friction between ice and the ship's flat bottom. These could be designed to diverge from the centre-line to impart a gentle sideways thrust to the ice. The effect would be to guide ice laterally at keel depth so that it would rise beneath unbroken ice outside the channel instead of alongside the hull. Less ice would reach the propellers. The idea of protecting the propellers of icebreakers is not new,[9] though it is particularly relevant to the problem of broad-beamed tankers in ice.

Apart from her great advantage of sheer mass, *Manhattan*'s performance in ice was related to her power/beam ration. It is a paradox of conventional icebreaker design that we position the broadest dimension at the level of maximum resistance to motion – the ice level. Tabulating the ratio of shaft horsepower to beam puts *Manhattan* well up in the major ice-breaking league. The units are meaningless but the comparison is significant:

Staten Island	16
John A. Macdonald	21
Manhattan	27
Louis S. St-Laurent	30

Clearly, anything that can be done to reduce width at the ice level in future designs would be handsomely rewarded in terms of ice-breaking ability, though it would sacrifice cargo capacity. Continued reduction in width would lead to a semi-submersible ship in which the volume lost by reducing width at the ice level could be replaced by greater width lower down. Carried a few steps further, we might conceive of a wasp-waisted tanker having a wide hull below sea level joined to a wide superstructure above through a longitudinal knife-edge structure at the ice level. I had sketched a design like this in 1969; a naval architect's version was published by Norway's Aker shipyard in 1975.[10]

Anything else that reduces drag should be considered. Air-bubbler or water-jet systems that separate ice from the hull are already fitted to some icebreakers, and durable low-friction paints have been developed.

However, some major questions first must be addressed. Could ice-breaking tankers compete economically with other forms of transport using alternative routes? Does Canada's sovereignty over the Arctic islands conflict with the rights of states to innocent passage through the archipelago? In Canada the question of jurisdiction is an emotive issue as well as one based on concern for the environment.[11] The government reaction to *Manhattan*'s voyage was pungent and predictable: years of argument about Canadian sovereignty in the Arctic.

The economic question had been answered months before the *Manhattan* set sail by the decision to build the pipeline from Prudhoe Bay to Valdez for onward carriage by conventional tanker. The diameter of the pipe was calculated to take all of the expected production of the oilfield. Indeed, the pipe sections had been delivered while *Manhattan* was in the Arctic. The consortium's haste to get their oil to market had overtaken the longer-term question of ice-breaking tankers.

Standard Oil, Humble's parent company, was one of those involved in the consortium. When the costs of laying the pipeline escalated alarmingly, Humble was forced to suspend its ice-breaking tanker studies, but stated that 'development work could be resumed at short notice if economic factors change or other circumstances warrant'.[12]

Tanker transport is not the only means of carrying Alaskan oil on sea routes. Since the Arctic pack ice makes surface navigation difficult and many a ship has been beset by ice, crushed, and finally sunk, why not go under the ice? Nuclear submarines have done so for years. USS *Seadragon* went safely through the Northwest Passage under the ice in 1960. As long ago as 1922, Stefansson wrote of submarines for carrying oil under the ice of the Arctic Ocean.[13] There is no doubt that it could be done. However, submarines are expensive and it has yet to be seen whether they can compete economically with alternative forms of transport. The Shipbuilding Division of General Dynamics Corporation offered to build submarine tankers of up to 300,000 tonnes,[14] but there were no takers.

I was left in no doubt that, in the long run, the *Manhattan* experiment would not be wasted. Analysis of the mountain of data took several years but eventually appeared in the form of proprietary reports. These yielded the essential design criteria for large ice-breaking tankers.

6.   CALM BENEATH THE WAVES (1971)

For those in peril on the sea – this time it is us

Lieutenant Commander Robin Whiteside, Royal Navy, came to the Scott Polar Research Institute in February 1971 to ask about Arctic sea ice. How thick was it? Was there a continuous sheet, or were there leads and open patches? What were the dangers to shipping?

Our curiosity was aroused. He revealed that he was First Lieutenant of Britain's first nuclear-powered submarine, HMS *Dreadnought*. There was a plan to venture under Arctic pack ice to gain experience of how to do it and where the ship could surface.[1] Another objective was to work at and around the ice edge with HMS *Oracle*, a diesel-powered submarine.

Whiteside had come to the right place. The Director of the Institute, Gordon Robin, was a World War II submariner; and I had spent four years studying pack ice in the Northwest Passage. A Ph.D student, Peter Wadhams, was working on the dynamics of drifting pack ice.

We talked for hours. Whiteside was defensive when asked where they were going – planned operations were secret. We knew that American submarines had been going to the North Pole and also right across the Arctic Ocean since 1957, so gently probed to see if such might be in mind for *Dreadnought*. 'That can't be ruled out,' he said – a diplomatic answer that could not be said to infringe the Official Secrets Act.

We pointed out how much could be learned about the ice from a submarine by using upward-looking echo sounders. The Americans had brought home thousands of kilometres of under-ice profiles but most of their data were still classified Secret. As far as scientists were concerned, that meant the results were as good as non-existent.

Having done our best to tell Whiteside all we knew, I saw him to the door. On the doorstep he paused, turned to me and said: 'You wouldn't like to come with us, would you?' My reply was delivered in a fraction of a second.

'Can't promise anything,' he said. 'We're off in a couple of weeks and you would have to have both British and American security clearance.'

'Why American?' I asked.

'Because the propulsion system of our submarine was built in the US and the technology is classified Secret. A condition of its sale to Britain – imposed by the Pentagon – was that all who sail in *Dreadnought* must have been cleared for security by the US Navy.'

My heart sank because a few years earlier I had spent 18 months in isolation with the Soviet Antarctic Expedition and that might compromise my position. People have been brainwashed in a shorter time. The Americans had nothing to gain from my participation, so why take the risk?

I endured two weeks of suspense. Meanwhile Gordon Robin negotiated with the navy for Peter Wadhams to join HMS *Oracle*. The news finally came that Wadhams and I had been cleared for security by the Royal Navy – but no word had come from the Americans.

It might mean that Wadhams could go and I could not. That was unthinkable – to me at least, but not to him. Luckily, all negative thoughts vanished with a telephone call from Whiteside.

'You are invited to join us at the nuclear submarine base at Faslane, on Gare Lock, on 19 February. Bring some cold-weather clothing.' I tried to elicit more details but all I got in response was 'I can't tell you more over the telephone.'

I had no official papers, no proof of identity, and no indication of who I should report to. But nothing venture, nothing win. So taking the night sleeper from Euston to Glasgow, I continued on a local train along the north bank of the Clyde to the tiny village of Garelochhead. Carrying my worldly goods in a large orange kitbag. I tried to explain myself to the armed guard on the gate of HMS *Neptune*, the submarine base.

It was the following morning before I was allowed on board. The sleek whaleback outline of *Dreadnought* lay next to the forbidding bulk of a Polaris missile submarine. There were sailors running back and forth loading last-minute stores. Whiteside welcomed me on board and I was ushered down a ladder to the control room. I had never been in a submarine before so it was all new to me. I recognized the periscope, the chart table, and a couple of control columns that looked as if they could have been taken from an aircraft. Beyond, every square inch of bulkhead was stacked with instrument and control panels. Men were everywhere, some looking surprised to see someone not in uniform.

All around the ship there was an air of great excitement – which I shared. I was told that we were heading for the Arctic Ocean, and if all went well, under the ice to the North Pole. So this was it – a Royal Navy first.

I was shown a small door at the after end of the control room. 'You may not go through there,' said my host firmly. A questioning glance from me brought the explanation: 'Your American security clearance did not come through, so we have to confine you to the forward end of the submarine.' So that was it. The engine room and nuclear reactor, while open to any of the hundred or so crew, was out of bounds to me. The door was never locked but I was honour bound to keep out.

Dreadnought, I learned, had been commissioned in 1963. A few years later, some cracks were found in the pressure hull and she had to be dry-

HMS *Dreadnought* under way

docked for a major refit. The refit was only completed in September 1970; so now, many things looked new and nobody spoke of cracks.

The ship had an illustrious name. The first *Dreadnought* was an Elizabethan warship launched in 1573; she fought against the Spanish Armada. The second was originally a Cromwellian frigate named *Torrington*, renamed *Dreadnought* when the monarchy was restored in 1660. The third *Dreadnought* fought both the French and the Spaniards. The fourth saw service in the West Indies. The fifth took part in the Battle of Trafalgar. The sixth was a shore station. The seventh was one of the first ironclad turret ships. The eighth was a battleship with ten 12-inch guns; then *Dreadnought* became the type name for battleships to follow.

The CO, Commander Alan Kennedy, was a handsome and friendly officer in his thirties – the sort of dashing young man one might expect to find on the front line of the Cold War. He explained that I was considered a guest of the wardroom but that submarines do not have spare officers' quarters for visitors.[2] I would be berthed on a bunk in the torpedo room – the fore-ends in naval terms. Since I well knew what a privilege it was to be on board at all, that was the least of my concerns.

There was one other civilian on board. Bill Thomson of the Admiralty Compass Observatory in Slough had come to supervise the operation of a Submarine Inertial Navigation System (SINS for short). The ship would depend on this bulky instrument to know where we were in the Arctic Ocean. The Navigator could take star sights with his sextant and there were also radio navigation aids but it was not known how far north they could be relied on.

I was invited up to the bridge as we departed on Exercise Sniff, as the voyage was to be known. At 0900 hours on 20 February the reactor 'went critical', lines were cast off, and we slid south down the loch past the little village of Rhu into the Firth of Clyde. Later, heaving on a gentle swell as we came out of the narrows, I was bid to watch as we increased speed and 'got on the step'. That 3,000 tonnes of steel could be made to rise up like a speedboat or a seaplane was a surprise to me, but rise we did. The large bow wave seemed to be resisting, but as the submarine was driven faster we began to overtake the wave until, forced under the centre of gravity, it suddenly heaved the ship upwards. A bow-up attitude gave way to an exhilarating surge of speed as our fore and aft trim became level. We were riding on a wave.

We passed a Russian spy ship – officially a trawler – anchored just outside the 3-mile limit of British territorial waters and in a position where she could monitor all comings and goings of the submarine fleet, surfaced or not. Such vessels are equipped with sonar so sensitive that it can record the unique 'sound signature' of every passing ship. Thus armed, Soviet warships could identify us anywhere. With our naval identification *S101* painted on the conning tower, today we made life easy for them.

I went below and some time later we dived. It was an eerie sensation as the ship pitched to a bow-down attitude and slipped beneath the waves. Only the depth gauge revealed what was going on. We levelled off at a depth of 90 metres as I came into the wardroom.

Dreadnought's track to the North Pole

The officers' quarters were just forward of the control room and up against the curving deckhead of the pressure hull. The Captain's modest cabin opened off the wardroom. By my standards, most meals were superb. The Captain sat at the head of the table with a signed portrait of the Queen

65

on the bulkhead behind. Officers wore collar and tie to meals, even the engineers who, moments before, had been seen in grimy overalls. Stewards waited at table; wine and spirits were freely available. A depth gauge was mounted over the steward's hatch so all could keep an eye on how deep we were. The atmosphere was relaxed.

In another sense, it was the atmosphere that would allow us to stay down for months – if necessary – in contrast to conventional submarines, which surface daily to breathe fresh air and to charge their batteries. So good is the 'air-scrubbing' in nuclear submarines that I never felt any symptoms of claustrophobia. Although all ranks are allowed to smoke as much as they wanted when off duty – and many did – the air seemed cleaner than in most shoreside wardrooms.

My sleeping quarters – the torpedo room – was the biggest compartment in the ship. I shared it with a dozen sailors. There were six 21-inch torpedo tubes at the forward end, the rest of the space being taken up by spare torpedoes, one above the other on racks. I spent the nights between two torpedoes. Awaking once with a start, my head came up against the torpedo above and I acquired a bruised forehead. It was an occupational hazard for supernumerary boffins.

I had served in the navy a quarter of a century before (1944–46) and was interested to note the changes. For the most part, I was conscious of a general relaxation in the once formal relations between officers and other ranks. No longer akin to masters and servants – they were now colleagues. Petty officers, in their halfway house, dealt easily with all ranks – though anyone rash enough to be cheeky would be brought up sharp. One day I found a chief petty officer being offered a drink at the wardroom table – unthinkable in my day. But in this situation, I thought, why not? The daily rum ration for the crew had been abolished in 1970 and in its place was a beer ration. As I passed the petty officers' quarters, smiling faces invited me in for a beer; I too was a colleague. Morale seemed high all round.

Our track north was to the east of Iceland and west of Spitsbergen. The only thing to punctuate the routine was a daily surfacing to send radio messages and to cross-check our position using Loran and Omega radio navigation systems. One day we came up in a hurricane and I had to hold on for dear life as the ship pitched and rolled heavily. But below 45 metres, waves were damped right out and there was no sensation of movement. The wardroom, being forward and far from machinery, was beautifully quiet. There was no vibration – a pencil would stand on end on the table. Since I am prone to seasickness on the surface in rough weather, I decided this was the ultimate form of happy sea travel.

It was a windowless world but what did it matter when there was nothing but waves to see on the surface? The total freedom from vibration was explained by the single propeller mounted aft of the aftermost part of the hull. Water thrown centrifugally from the blades was not hitting any part of the hull – as it does on surface ships.

I read various manuals in the wardroom to get a feel for how things worked. People talk of nuclear-powered ships as if their engines were special. In fact *Dreadnought* was a steamship driven by steam turbines no different, in principle, from those that had powered most surface ships for more than a century before diesel propulsion became the norm. The nuclear reactor provided heat – as had coal and oil before. The critical difference in a submarine is that steam cannot escape to a funnel. A leak in the primary pressurized-water cooling system around the reactor could result in a dangerous dose of radiation for any of the crew near it. A leak from the secondary (non-radioactive) steam circuit driving the turbines could scald anyone on duty in the engine room. But nothing leaked, and we were driven along by the turbine's 15,000 shaft horsepower.

Clearly, a thousand things could go wrong, and the crew had to be trained to react quickly and correctly to any of them. Instead of having sleepless nights thinking about it, I put my life in their hands just like anyone who flies the airlines. I had no choice.

There were three upward-looking echo sounders and I volunteered to monitor them to make sure that we brought home a continuous profile of the underside of the ice. Time marks had to be pencilled on the record and every few hours one of the paper rolls came to an end and had to be replaced. Before putting it away, I wrote down the starting date and time, and its finishing time.

Nowhere was out of bounds to me except the engine room, so I wandered round the accommodation areas to talk with the inhabitants. I never asked questions about things that I knew to be secret, such as the top speed of the submarine. But for that, one only has to look in the pages of *Jane's Fighting Ships* to find their best guess: 30 knots dived. We never approached that speed because the engineers knew that the harder a ship is driven, the more likely it is to precipitate some failure in the propulsion system. That is not what we wanted while confined beneath a canopy of ice.

From time to time we rose up to periscope depth. The Officer of the Watch (OOW) rapidly scanned the horizon to see whether any surface ships were around. But all we saw was a grey horizon and waves splashing over the outer lens. We crossed the Arctic Circle on 22 February. King Neptune and his followers initiated all the novices on board to the honourable company of 'Bluenoses' amid much hilarity and cold water.

Two days out of Scottish waters we approached and then contacted *Oracle*, but two more days passed before we found the first belts of pack ice west of Spitsbergen. *Dreadnought* then spent a full day playing hide-and-seek with the other submarine. The significance of experiments at the ice edge is that the noise of ice floes grinding against each other can make it hard to interpret both active sonar (echo-location) or passive sonar (listening for engine noise). Difficult or not, it was important to know at what distance we could detect the presence of another submarine.

Leaving *Oracle* to continue her own experiments without us, we headed for Prins Karls Forland, a rugged island off Spitsbergen. Passing the island 22 kilometres to starboard, we saw only the splashed image of a distant mountain through the periscope. On reaching latitude 80°N – 1,100 kilometres from the North Pole – we found an outlier of the main Arctic Ocean pack ice. I could watch the upward sounders trace the rugged profile of the ice floes above. Their downward-projecting ridges are known as ice keels, the counterpart of the characteristic pressure ridges seen from above. The difference is that an ice keel, to support the ridge above it, may extend downwards up to ten times the height of its ridge. For this reason ice keels can be a hazard to submarines.

The Captain was inclined to go deep in case we encountered an iceberg. Being responsible for the safety of his ship and all who sailed in her, he was right. My interest was in going shallower, not because I wanted to hazard the ship but because it would yield better under-ice records. In principle, each of the echo sounders was emitting a cone of sound upwards. The deeper we were, the more generalized was the recorded profile of the ice above. Going shallower would 'illuminate' a smaller area of the ice and show more detail – which is what I wanted. I explained that icebergs had never been seen on our intended track. That was true but had I been more honest, I would have added that there was seldom anyone in these waters to report an iceberg if there had been one.

Most of the time the keel of the submarine was kept 85 metres below sea level. Being 18 metres high from keel to fin, the top of the fin (the conning tower) was 68 metres below sea level. This gave a reasonable safety factor – though I knew that 45-metre ice keels had been reported in other parts of the Arctic Ocean.

The last day of February was Sunday, the day for 'Captain's Rounds' and for church. The whole ship was made spotlessly clean and tidy – decks scrubbed and dusters deployed in the most unlikely places. Like everyone else, I made sure that I was shaved, my bunk was tidy, shoes clean, tie straight – then stood at attention beside my torpedo as the Captain's procession passed by.

The church service, led by the Captain, was something I will long remember. It consisted of lessons, prayers and hymns, ending with the sailor's hymn:

> *Eternal Father, strong to save,*
> *Whose arm doth bind the restless wave,*
> *Who bidd'st the mighty ocean deep*
> *Its own appointed limits keep:*

O hear us when we cry to thee
For those in peril on the sea.

I had a lump in my throat and there was scarcely a dry eye in the house. I think that the sailors were slightly embarrassed by allowing emotion to get the better of them but I knew what they had in mind. This time, it was *we* who were in peril on the sea – more so than any other ship in the Royal Navy.

7.   THE NORTH POLE (1971 continued)

The first British ship at 90 degrees North

It was only 12 years since the Captain of USS *Nautilus*, heading for the North Pole, felt that he was undertaking 'the most thrilling and adventurous cruise any sailor ever embarked upon'.[1] The Captain of USS *Skate* felt that he was taking 'the gravest risks ever asked of a peacetime crew'.[2] Some of their tales had been hair-raising. They wrote of surprise encounters with icebergs whose vast bulk reached well below the submarine's cruising depth. On the Alaskan side of the Arctic Ocean, the Bering Strait and Chukchi Sea were notorious among submariners because the sea is so shallow that in order to clear the ice above, there is almost no clearance below. I could all too easily imagine the jolting crash on hitting the 'floor' before rebounding to hit the 'ceiling'.

Nobody in *Dreadnought* had ever surfaced through ice, but the American narratives told how the technique had been developed and what had gone wrong along the way. I saw that just reading these had lifted some weight off the minds of our officers. The theory was straightforward – but could we learn to do it in practice and come home unscathed?

I was asked to give a talk to the crew about ice in the Arctic Ocean. Some of them were apprehensive about possible difficulties in surfacing. The *Nautilus* and *Skate* voyages we had read about had both been in summer. Now it was winter – would there be any openings at all? Luckily I knew from later voyages of US submarines that openings had always been found within a few hours of starting a search. Not all the sailors were reassured. It dawned on me that for submariners, the ease of surfacing at sea to deal with emergencies compensated – to some extent – for the feeling of being trapped in a sealed cylinder far below the surface. Their greatest fear was fire. My assurance that openings could be found in the ice was – they all knew – cold comfort in relation to the speed and survivability of a fire.

Kennedy was anxious to relieve the anxiety of some of the crew by showing that we could surface through the ice. So after breakfast on the last day of February there was intense interest in the upward sounder records. If the OOW reported thin ice above us for more than the length of the submarine (81 metres), we would double back and cross the opening at an angle of 60 degrees to the first crossing. This was necessary to find whether the ship had been unwittingly travelling along a narrow lead instead of an open patch large enough to surface through. If the second pass indicated an opening of a reasonable size, a third pass would be made at an angle to that. If it was finally concluded that the area was suitable, the ship would be slowed and navigated – by dead reckoning – back to a point presumed to be in the middle of the opening. Scientists use the Russian term 'polynya' to describe an opening in pack ice, but an opening covered by thin ice is called an ice skylight. We had found a skylight. A buzz of activity followed the announcement of 'Hands to diving stations – stand by to surface through ice'.

A submarine controls its depth by means of hydroplanes, one pair forward and another aft. In principle, aircraft are controlled the same way, though with a single pair of elevators aft. But there the similarity ends. Whereas aircraft must maintain flying speed, submarines can stop dead in the water. There are complications to what is called neutral buoyancy. Without water flowing over them, the hydroplanes are useless. In neutral buoyancy, if one end of the submarine is marginally heavier than the other, there is nothing to prevent the bow progressively pitching down – or up, as the case may be. There was a large spirit level in the control room mounted fore and aft to show at what pitch angle we were. As we came to a standstill and began, albeit slowly, to pitch down, there were several anxious pairs of eyes on the bubble.

I was wondering what technological fix could deal with this situation. As it turned out, technology was not the solution. A 'trimming party' of five sailors was directed to run to a given position, in this case aft of the centre of gravity. In relation to our submerged mass of 4,000 tonnes, this was like putting a feather on one side of a kitchen balance – but it worked. The trimming party was moved in one direction or the other to keep the submarine level.

The next stage was to lighten ship by pumping out water ballast. Once the ship was more or less level, eyes darted back and forth to the depth gauge. Tension mounted in the control room. By subtracting the echo-sounder depth from the pressure depth, we had a rough idea of the ice thickness above. Here it was 0.3 metres. We came up steadily at the rate of 5 metres per minute, with the bubble slightly bow up.

'Up periscope!' at 67 metres. The water was clear, with good visibility. Floodlights mounted on top of the casing showed vague shapes above. A camera was pushed up against the periscope lens to record the sight. The rate of ascent slowed to 2 metres and then increased to 6 metres per minute. 'I have seen some goddamned funny things in my life but this thing takes the biscuit,' said Kennedy for all to hear. 'The ice looks light and I can see some clear water.'

At 31 metres' depth the ice was 14 metres above the fin. After three minutes there was a slight shock as the fin hit the ice and we stopped rising. Evidently we were not surfaced, so the Captain called for a one-second 'puff' of compressed air to expel water from the ballast tanks. Nothing happened. Another puff, then another – all eyes were on the depth gauge. More puffs and the fin saw the light of day. We halted at a keel depth of 11 metres, the top of the fin 7 metres above sea level.

The lower 'lid' (hatch) to the conning tower was opened and men shinned up the ladder to open the upper lid. Ice on the outside lens of the periscope obscured the view, so a heater was switched on to melt it. We had surfaced through slushy ice and the outside air temperature was –8°C. Those not on watch scrambled up the ladder to exit through a side door of the fin. The top of the casing was slippery and I trod carefully because there was nothing between me and ice-water crowding in from both sides. A fall down the sloping side could be disastrous. The First Lieutenant called for a line to be brought up in case someone did slip. We were shivering but felt triumphant.

Dreadnought spent two hours on the surface. When the air temperature dropped to –16°C, it was time to see whether or not any topside valves had frozen. Luckily they had not. Everything below sea level on a submarine is kept warm – in a manner of speaking – by the sea, which in its liquid state can never go below –2°C. After testing the hydroplanes, down we went, this time releasing air from the ballast tanks and taking on sea water. With neutral buoyancy at cruise depth, the navigator set course for the North Pole.

At intervals we slowed to 3 or 4 knots to check the fore and aft trim, and to cycle the ballast tank valves to make sure that all were functioning. The ice above became more and more rugged. There was a good deal of level ice about 2 metres thick punctuated by ice keels projecting as much as 24 metres down. Having sailed in icebreakers, I was able to picture the scene above – hummocked ice floes with blocks of ice thrown up to form jumbled ridges 3 metres high.

At 82°N the gyro compasses were released from their normal north-seeking mode to become free gyros. At some point in approaching the North Pole, gyros can no longer detect the rotation of the earth – the Coriolis force. However, free gyros on frictionless bearings should be able to maintain their set alignment and still be useful for navigation. The only problem is that there is no way – apart from star sights or other means of direction-finding – to be certain that they are still aligned.

From this point on, the sun was below the horizon. It would not rise again until we passed the same latitude southbound. The periscope became useless except for an occasional faint loom of twilight when surfacing. We had entered the polar night.

I watched the echo sounders for hours on end because they were full of surprises. You never knew what was coming from the unfolding profile above – gigantic ice keels, kilometres of level ice, or polynyas big enough to take six submarines. I saw the reassurance on the officers' faces every time we passed a good-sized skylight. Like pilots in the sky above, they were alert to a host of possible emergencies and how to react to each one.

At noon on 1 March we were at 85°N and found a place to attempt surfacing. Doubling back as before to establish the size of the polynya, we found that it looked just large enough, provided that we did not drift while coming up. That was easier said than done, because the normal ship's log was not designed to measure speeds that would be negligible in other circumstances. The crew were told not to move about the submarine – it upset the trim. Here too the echo sounders indicated about 0.3 metres of ice above, but thicker ice close to the bow and stern.

There was a small bump as we hit the ceiling, with cracking noises on breaking through. The submarine came to rest with a 3-degree bow-down attitude and a 3-degree list to starboard. This was ominous because we had been nicely in trim while rising. Something must be upsetting the trim. The first people on the bridge reported fog, a gusty wind, snow falling, and an air temperature of –21°C. Worse than that, there was a 2-metre-thick ice floe lifted up on the bow and on the starboard fore plane. It was bad enough having a weight of tonnes bearing down on the hydroplane, but the bow-down attitude meant that the single propeller, unprotected from any angle, might be bearing on the underside of an ice floe aft.

If the main shaft should be damaged, we could be disabled – with or without functioning control surfaces. The Captain decided that our situation was unhealthy. The call 'Hands to diving stations' was followed by venting the main ballast. As we submerged, the 3-degree bow-down angle gave way to a 7-degree bow-up angle. There must be ice under the bow, and we were stuck to it. With the propeller now down and clear, it was turned dead slow astern to pull us away from the ice at the bow. At the same time ballast was pumped from aft forward in an attempt to regain trim. The trimming parties too were sent forward.

Minutes passed before we dropped away, free at last but with some shattered nerves. Luckily no damage was detected. We had learned that, with only three echo sounders, personnel in the control room are blind to what may be overhead in the waters fore and aft of the cone that each instrument 'illuminates'. American submarines had learned this lesson the hard way and were equipped with seven sounders, virtually ruling out blind spots.

The following day we surfaced at 87°N. We knew the sky would be dark – not a trace of light came through the ice. But the floodlights lit the underside of the ice canopy when the periscope was 12 metres beneath. When the OOW climbed to the bridge, he found that we were in a lead but that the bow was touching heavy ice. It was a clear, cold day (–24°C) with a 20-knot wind. A 'wind-chill' diagram showed that with this combination the effective temperature on bare skin was equivalent to –48°C.[3]

We spent eight hours on the surface, testing various instruments and running a diesel generator to charge batteries. The navigator attempted star sights for position, and a good 'Loran-C' fix showed that we were right on track. Later in the day the wind increased and the temperature fell to –33°C.

The OOW triumphantly announced that the wind-chill diagram yielded an effective temperature of –61 °C. Nobody hung around topside. Some saw the tell-tale white patch on the cheek of a colleague – frostbite. More than once I felt the pinprick sensation as my cheek froze. A warm hand can thaw it – though not for long under these conditions.

Dreadnought's designer had not planned for this environment. Various fittings on the bridge iced up, the periscope bearings froze, and several ballast tank valves would not open. But enough did, and after the call to diving stations, we fell away at 18 metres per minute. Everything thawed in the (relatively) warm sea.

There was plenty of entertainment on board. Films were shown in each mess every day. Backgammon was popular in the wardroom. 'Uckers', the navy's special version of Ludo, was also much in evidence. There was a ship's newspaper, *Dreadnought Express*, with news bulletins, poems, puzzles, cartoons and articles – even some that I wrote.

We reached the North Pole at breakfast time on 3 March. However carefully we manoeuvred to get the SINS to read exactly 90°N, it would not do so, for the good reason that the first digit included only numbers from one to eight. However, I do have a photograph of the instrument reading 89°59.9'N. If that was the truth, we were 200 metres from the pole – good enough for me.

I rarely brought out my camera, realizing that I was surrounded by instruments of varying degrees of secrecy. When I did, at least two pairs of eyes ensured that I pointed it only in approved directions. I was allowed to photograph the planesmen at the hydroplane and steering console only after the most sensitive instruments had been covered. Shields were standard equipment for occasions when visitors were coming aboard in harbour. Outside, of course, I could photograph to my heart's content; but in the blackness of a polar night, there was not much to see.

Everyone was keen to surface at the pole and to go for a walk on the ice. It took hours of searching, and one failed attempt, to find a good spot through which to come up. Having found an ice skylight with 30–60 centimetres of ice, we stopped, trimmed, puffed, and then thumped the ceiling at an upward speed of 5 metres per minute. We heard no sound. The 'sound room', however, with its sensitive listening devices, reported three cracks from the ice.

The first man to breathe polar air found the small door in the side of the tower jammed by ice, and had to fight his way out. When I came up, no part of the hull was visible. Instead of the black casing stretching fore and aft, there was only a low ridge of smooth ice lifted by the rising hull. No water could be seen anywhere, so we could safely step down to the smooth ice on either side.

The sky was clear, there was a three-quarter moon, air temperature –37°C, and a 20-knot gusting wind. The OOW's wind-chill diagram made that –65°C, and we felt as much. Thirty seconds without gloves to fiddle with a camera and I was in pain. It took minutes of violent exercise to thaw the fingers. The crew quickly resorted to football. Others unfurled the white ensign and a 6-metre-long banner reading HMS DREADNOUGHT NORTH POLE 3.3.71. But floodlights had to be deployed on the ice before anyone could record the occasion. We felt elated, proud of the navy, and proud of our country.

In the direction of Canada, a faint light suffused the sky with a band of orange on the horizon. It was the beginning of the Arctic dawn – the sun was just 7 degrees below the horizon. Sunrise – the one and only sunrise of the year at this spot – would occur two and a half weeks after our visit. Captain Kennedy came up to me on the ice and, to my surprise and delight, presented me with a bronze ship's crest mounted on oak. The motto on it admonished me to FEAR GOD AND DREAD NOUGHT. I will treasure it always. The emotion of the occasion brought tears to my eyes – but they froze as fast as they trickled down my cheek.

Walking off to investigate the size of the skylight, I came to hummocks 2–3 metres high. The level ice was 22 centimetres thick with 3 centimetres of snow on top, but it cracked ominously underfoot. Finding myself alone, I recalled the moment in Fury and Hecla Strait long ago when I came upon the large fresh footprints of a polar bear. Now there might be one attracted by the rare commotion and about to pounce from behind a hummock. I retreated, glancing over my shoulder. However, I need not have worried – the First Lieutenant had posted a rifleman on the bridge and I was being watched.

Dinner in the wardroom was a memorable occasion: the Captain at the head of the polished table, white uniform shirts, black ties, mixed grill before us, wine, fresh fruit and a festive atmosphere. There was jubilation at having reached the climax of our voyage. I could not help comparing our feast with the meagre rations of explorers who had struggled over the ice to this place. Ours was a life of luxury.

Robin Whiteside had carved a rubber stamp to record our attainment of the pole, and set about franking hundreds of letters and postcards to be mailed at the end of the voyage.

The Captain extended the festive atmosphere by holding the ceremony of 'Requestmen'. He advanced four men to the next higher rating, awarded two good conduct badges and two specialist qualifications. Those men will long remember the time and the place of their awards.

After eight hours on the surface, we dived and set course south. That was all we could do, because any and every direction from the North Pole is south. The trick is to find the right direction – otherwise we might be heading for Russia, Alaska, Canada or Greenland. The Captain had to say what meridian or longitude to follow. He set off down 0° – the Greenwich meridian – though later various courses kept us between 1°E and 7°E.

One day there was a practice torpedo-firing without a torpedo. I was curious how this could be, so went below to watch. Short of loading a torpedo into one of the tubes, the drill was done as if it had been. Sea water was allowed in, and on a command from the control room, 'Fire One!', there was a loud but short-lived hiss and off went the imaginary torpedo. It was an 'air shot', with compressed air driving out the contents of the torpedo tube – as it would have done with a real torpedo.

The thickest ice, and the most closely spaced ice keels, were not at the pole but between 88°N and 86°N. *Dreadnought* encountered several ice keels down to 24 metres and at least one at 30 metres.[4] Had we found icebergs – which go much deeper – my name would have been mud. More likely, we could still be posted as 'Missing – presumed lost at sea.'

We surfaced through ice twice more before leaving the polar pack early on 7 March at latitude 79°N. The Captain ordered a practice 'Reactor Scram'. This involves shutting down the reactor as in an emergency. Nobody wants to be near a malfunctioning nuclear reactor immersed in superheated steam – the word *scram* makes abundantly clear what one would have in mind if it happened. Since nobody can escape from a submarine in a hurry, there were elaborate drill sequences that had become second nature to the crew. In this case, all the action was in the after part of the submarine, where I was not allowed. The engines shut down and we drifted to a stop, trimmed, and ballasted to stay at the same depth. After a time, the reactor 'went critical' and we proceeded as before.

Visiting the sound room one day, I was fascinated to hear coming from the sea, clicks, squeals and whistles. Asking the sonar operator what it was, he dismissed the sound as 'biological noises'. He was trained to listen for other noises. However, I knew biologists would have been fascinated to have a recording, because the uneven distribution of whales, seals and bears in the Arctic Ocean is still very little understood.

Dreadnought, of course, was making her own noises, chiefly from the propeller. We knew that the Russians were interested in what we were up to and would be listening – somewhere – to track our progress as best they could. Having our own reasons for not wanting to encounter a Soviet submarine – particularly head-on – we kept our active sonar pinging away throughout the voyage.

From time to time, water samples were taken for chemical analysis. Sea temperatures and the velocity of sound were recorded every hour because all measurements of underwater distance depend on knowing the velocity of sound. That varies with salinity and sea temperature, but much of the time it hovered around 1,440 metres per second. Soundings of sea depth in the Arctic Ocean showed steep underwater mountains, ridges, and abyssal depths – some down to 4,300 metres.

The homeward voyage was something of an anticlimax – offset, however, by the warm glow of achievement. I was invited to take the hydroplane controls for a time and found it not unlike handling an aircraft. With my eyes glued to the artificial horizon, it was not difficult to maintain depth to within 1 metre.

On the morning of 11 March we slid quietly into *Dreadnought*'s berth at Faslane, to be met and congratulated by FOSM (the Flag Officer Submarines), Vice Admiral Sir John Roxburgh. We had steamed 9,600 kilometres in 19 days and surfaced six times through ice. The only damage sustained consisted of two small dents in the bow. While under the ice, two members of the ship's company became fathers, and a 19-year-old ordinary seaman said he was the youngest Welshman ever to visit the North Pole.

At this stage my interest was to analyse the upward echo-sounder records in terms of ice thickness, frequency, the size of ice keels, floe sizes, and the occurrence of polynyas. I asked Kennedy whether the records were to carry a security classification – as do American submarine data. I pointed out that since the Americans and also the Russians must by now have thousands of kilometres of Arctic ice profiles, why classify ours? Surely there were no security implications if these were made public? He thought a bit and replied, 'In cases like this where there is no precedent in the Royal Navy on how to classify a new class of documents, it is left to the Commanding Officer to rule on each case.' My eyes lit up hopefully and he said, 'Let's call them unclassified!' With that I threw a full set of records into a mailbag and returned to Cambridge.

In May 1971 there was an international conference on sea ice held in Reykjavík, Iceland, and I was able to present an analysis of the *Dreadnought* data.[5] The conference was attended by all the leading scientists in the field, including Dr Waldo Lyon of the US Naval Undersea Research and Development Center, and Walt Wittmann of the US Naval Oceanographic Office, both of whom had much experience of submarine ice profiling. They thanked me for making our records public because it would help them to argue that their own data should be declassified.

A later and more exhaustive analysis was completed by Elizabeth Williams of the Scott Polar Research Institute, together with Gordon Robin and me.[6] It led to a long period of collaboration between the Royal Navy and the Scott Polar Research Institute on subsequent under-ice voyages.

For me, *Dreadnought*'s North Pole voyage remains a very happy memory.

Ralph Maybourn inside *Manhattan*'s massively ice-strengthened bow

Consolidated pack ice in Viscount Melville Sound

The new Canadian icebreaker *Louis S St-Laurent*

Edward Platt with the golden barrel of Prudhoe Bay crude oil, the first commercial cargo carried through the Northwest Passage

We surfaced in twilight at 87°N

HMS *Dreadnought* at the North Pole, 3 March 1971

Celebrating at the North Pole. Captain Kennedy sits at the head of the table

The Comet at Thule Air Base. I am half way up the steps with sunglasses

8.   COMET OVER GREENLAND (1971)

How to fly a jet airliner

Two months after returning from the North Pole a strange request came from the Aeroplane and Armament Experimental Establishment at Boscombe Down in Wiltshire. They wanted to do some low flying over Greenland to test new instruments. Lacking staff experienced in Arctic matters, they needed someone to tell them what kind of terrain they were flying over. It was the sort of opportunity that took me a split second to think about. It promised a free-of-charge six-day sightseeing holiday with spectacular scenery

Gordon Robin successfully negotiated to get Peter Wadhams on the flight. Peter was studying the penetration of ocean waves into pack ice and we were to pass over thousands of kilometres of various kinds of sea ice. I wondered how our flight could contribute to his study until it was explained that the aircraft was fitted with a laser altimeter. This device could determine our height over the ice to an accuracy of a few centimetres. If we were flying level, it was precise enough to measure the changing wavelength and wave height as ocean waves become attenuated by the ice. A few weeks earlier, Peter had been measuring the same thing from under the ice in HMS *Oracle*.

The prospect became more exciting when we were told that the aircraft was a Comet-4C, a four-engined jet airliner developed from earlier Comets, several of which had exploded in mid-air in the 1952–54 period. The original Comet was the world's first commercial jet airliner.

The flight had two objectives. One was to compare two new types of radar altimeter over snow and ice. The second was to study which of several navigation systems would perform satisfactorily in high latitudes. Our security clearances from the submarine trips were still valid. However, as all the instruments to be tested were off-the-shelf commercial products, the operation had a low security rating.

Peter was in Iceland when the time came for the flight but as the plan was to refuel at Keflavík, the NATO airfield near Reykjavík, he was instructed to meet us there. I was asked to report to 'Hangar E1' at Boscombe Down at 0900 on Friday, 21 May 1971.

Although we were planning to fly as far as 84°N, we needed very little extra clothing because most of the time we would be confined to an air-conditioned cabin as comfortable as any civil airliner.

Appearing in hangar E1 at the appointed time, I was greeted by Squadron Leader A.M. Chandler, Officer in Charge of the Arctic flight, and given coffee. As a reminder that this was not a normal airline waiting room, there was a poster on the wall reading:

> Aviation itself is not inherently dangerous, but like the sea, it is terribly unforgiving of carelessness, neglect, or incompetence.

With a six-man RAF aircrew, I did not anticipate problems of that sort. The Comet was outside the door, brightly painted in white, red and grey. Her name was *Canopus*, RAF registration XS-235. I have always thought that the Comet was one of the most beautiful aircraft ever built – sleek, proud, and unencumbered by awkward engine pods like most airliners. Two streamlined fuel tanks faired into the outer wing just added to the aura of power and speed.

Climbing out over the rolling Wiltshire countryside, I went to explore the aircraft. In command was Flight Lieutenant John Foreman and in the co-pilot's seat was Flight Lieutenant W.C. Wackett, known to his colleagues as Swash. In what would be the passenger cabin in civil aircraft, I found the navigator, radio operator, air engineer and air loadmaster, together with one officer and six senior NCOs in charge of ground servicing. In addition, there were two squadron leaders, eight instrument specialists, and a photographer to look after a vertically mounted survey camera. Service personnel were in uniform and the civilian staff wore collar and tie.

The seats had headrests and were rather more comfortable than airline seats. Alternate rows had been removed and the spaces filled with instrument racks. I spent the three-hour flight to Iceland getting to know everyone who was not too intent on watching dials. All told, we had on board 24 men and one woman. She was Jackie Lampard and her job was to operate a computer that compared and evaluated every bit of navigational information supplied by a plethora of instruments. I recognized Omega, Loran, doppler and inertial navigation systems. As if those were not enough, for good measure there were two astro-trackers, direction finder (ADF), Omnidirectional radio range (VOR), and Tacan systems. We were not worried about getting lost.

Peter Wadhams came on board at Keflavík, and after refuelling, we headed for Thule Air Base on the north-west coast of Greenland. At an altitude of 8,000 metres over Denmark Strait, Peter spotted the ice below and asked the pilot to fly lower so that we could get the best out of the laser altimeter. The crew were not averse to low flying but Foreman explained that it would cost them one tonne of fuel to climb back up to cruising altitude. This gentle rebuff underscored Peter's status as a hitch-hiker. I was in the same boat, so avoided asking for anything.

Thule is a US Air Force base, part of North America's first line of defence

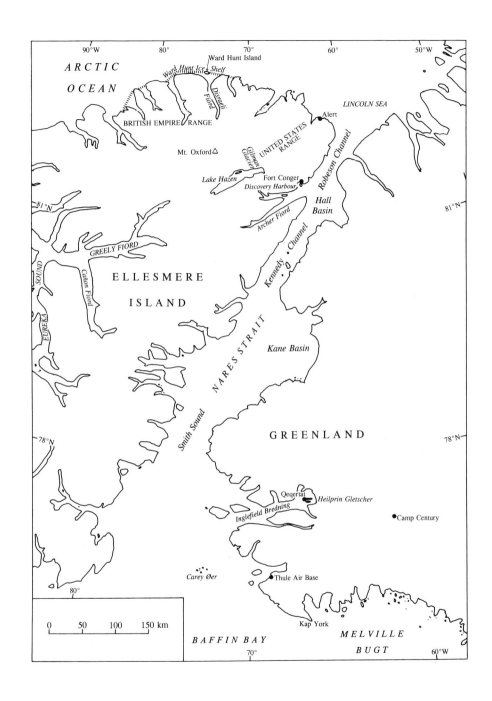

Map 6 Ellesmere Island and Northwest Greenland

81

against attack from the north. Its single runway occupies a glacial outwash valley carrying meltwater from the vast inland ice sheet that we could see in the distance. On final approach we passed rows of accommodation blocks that at one stage were said to have housed 10,000 men. On the ground it was a calm, almost cloudless day, and we relished the dry, clean air only a few degrees below freezing. The place was tidy enough but it seemed almost deserted.

A bus drove us to an empty housing block and the driver showed us where to find the mess. We enjoyed a good dinner, watched some blue movies and fell into bed. The next morning I witnessed a tiny event that seemed to epitomize the gulf between American and European norms. We were queuing in the chow line for breakfast. The man in front of me asked for a fried egg 'sunny side up'. As the egg plopped onto the griddle, the yolk broke. In a split second the cook had swept it into the dustbin and taken another. I was next in line and asked for two scrambled. The cook took two eggs and promptly broke the yolks with his spatula. It had not occurred to him to save the egg from the last man for the next one asking for scrambled.

Wasting one egg would not bankrupt the US Air Force. My point was that the egg – and everything else on the menu – had been airfreighted 4,500 kilometres from an air base in New Jersey. Later I was to find many more examples of extravagance – but none so poignant as the egg.

Dreading the first half-hour in a cold-soaked flying machine, we were delighted to find the Comet in a heated hangar, fully fuelled and ready to go. All we had to do was jump on board and fire up the engines. The object of the first flight was to reconnoitre the neighbourhood while trying out the altimeters on different types of terrain. As soon as we were off the ground I was summoned to the cockpit to sit in the jump seat between the pilots. I could talk to the pilots on the intercom but my job was to do a running commentary into a tape recorder. The pilots could see the terrain we were flying over but the tape would come into its own afterwards when synchronized with the record of the instruments. I had the best seat in the house.

Most of the time we flew at a height of 150 metres over the ground. We cruised up and down glaciers, over the inland ice sheet, icebergs, pack ice and open sea. I noted whether we were over bare ground, soft snow, hard snow, superimposed ice, bare ice, icebergs or ice floes. I had to estimate the depth of snow, the thickness of glaciers and the concentration of ice floes. Above all I had to mark the exact moment when we passed from one type of terrain to another. I felt sorry for the poor instrument specialists who could listen to the running commentary but were too busy to look at the beauty outside.

The following day we went far inland over the ice sheet to find areas with deep, soft, dry snow. It is under these conditions that some radio altimeters have been known to give false readings because the signal is reflected not from the snow surface but instead from a deeper layer. In the worst cases, aircraft have crashed because the altimeter was reporting their height over the reflecting layer instead of over the surface. The laser altimeter was recording the true height, but up front we had no read-out from it, so judged a safe height by eye. In the event, both of the new altimeters reflected from the snow surface and proved reliable over every kind of terrain.

One of the instruments had a broad beam whereas the other was more focused. This gave an interesting result as we glided down badly crevassed glaciers. The narrow beam responded so rapidly that at one moment it was measuring our height above the surface and a millisecond later it was plumbing the bottom of a crevasse. The difference told us the depth of the crevasses. In practice it all happened too fast for the human eye – the discovery came later on examining the high-speed recorder traces.

We managed two flights on 25 May. Working our way up the coast, we turned east into Inglefield Bredning and headed towards Heilprin Gletscher. We were flying 100 metres over the sea, and without looking closely at the map, found ourselves heading right at the small Eskimo village of Qeqertat. We were across it in a flash, too fast even for anyone to come out and wave or – more likely – curse. I have wondered ever since what they thought of a hedge-hopping four-engined jet bursting across their peaceful settlement. We climbed one glacier and came down another into Nares Strait.

Before the flight I had been speaking with Arnold Baker, one of the altimeter specialists. In his spare time he was a private pilot who enjoyed flying a Tiger Moth, photographing crop patterns from above and sometimes finding traces of Stone Age settlements and Roman villas. I told him of my experiences as a private pilot in the US and Canada. During the second flight of the day he had a quiet word with the Captain of the aircraft. The first I knew of it was when Swash Wackett climbed out of the co-pilot's seat and motioned me to take his place. There was a pause during which I scanned the instrument panel, then Foreman said: 'Would you like to have a go?' He did not have to wait for an answer. The aircraft was fingertip light on the controls and perfectly trimmed. Foreman said quietly: 'Hold her at five hundred feet.'[1] After some minutes during which I held tolerably close to this altitude, he said: 'You can come down to two hundred feet.'

We were cruising at 320 knots (170 metres per second). At this height, the sensation of speed was astonishing. It was a lesson in perspective. Icebergs and ice floes, first sighted at the vanishing point ahead, raced towards us along diverging lines of perspective to flash past on either side. It was the thrill of a lifetime. Diverting my gaze to admire the scenery, I came down at one point to 50 metres but recovered, I hoped, unobserved.

After 30 minutes of what to me was high drama, Foreman broke the spell with the words: 'I have control.' I relaxed in my seat and thanked him profusely for the experience. I wondered whether any other pilot in history had his first lesson in flying jet airliners at an altitude of 60 metres.

Skimming the surface of a glacier

Just then I saw something ahead that was not diverging from the vanishing point. A second elapsed before there was an almighty crash. It sounded like an explosion and sent a shock wave throughout the airframe. My reflex action on seeing three specks racing towards us had been to put my hand up to shield my face. Now I saw blood and small feathers splashed over the windscreen in line with my head. Without the armoured glass, I could have been decapitated. The gulls had no time to suffer.

Foreman said nothing but opened the throttles and climbed away to put some space between us and the ground in case the engines had swallowed some of the birds. At this point I felt it prudent to give up my seat to have the co-pilot return to his post. Everyone on board had felt the impact and wondered about the cause. A glance at the windscreen and my deathly white face told the story – 'bird strike!'

On taxiing up to the hangar, a crowd gathered to inspect the nose of the Comet. There were three more bloody patches, one on the nose and two on the port wing root close to the No 2. engine intake. Luckily none had been ingested, and the dents in the skin were negligible. I thought of the Comet-1 disasters and how their story must have begun the same way – with a bang.

There was many a muted sigh of relief when the senior engineer reported that we were still airworthy.

There was one kind of terrain that we had not flown over. Ice shelves are floating glaciers that, in Antarctica, extend over more than a million square kilometres and range in thickness up to nearly 2 kilometres. In the Arctic there are only a few small and thin ice shelves. The largest of them – Ward Hunt Ice Shelf – is on the northern coast of Ellesmere Island. So it was agreed that the last flight of our Arctic series would take us into Canadian airspace.

Cruising north between Greenland and Canada, we entered Smith Sound, then Kane Basin and Kennedy Channel. Here I suggested turning left to take in some mountain scenery and, as a bonus, to test the altimeters over lake ice – something we had not done before. The area was rich in history and I felt privileged to see it. We flew over Discovery Harbour, where HMS *Discovery* had wintered in 1875–76. Captain George Nares, RN, had been instructed to explore to the highest latitude possible. Five years later the site was occupied by an American expedition under Lieutenant A.W. Greely of the US Army. Subsequently the place was used, on and off, for ten years from 1899 by Robert E. Peary in the course of his attempts to reach the North Pole.

Lake Hazen on Ellesmere Island is a fresh-water lake 76 kilometres long and up to 12 kilometres wide. The early explorers spent days sledging in from the coast – we covered the ground in nine minutes. I had never seen the lake but had read about it. There was an abandoned research station on the north shore and as we passed by I saw two small huts. When the altimeter tests over the ice-covered lake were done, we turned north and climbed over Gilman Glacier. It was not on the chart and the crew was impressed that I knew its name. In the event I could have got away with calling it the Timbuktu Glacier.

Now we headed north between the British Empire Range and the United States Range – very grand names for not particularly grand mountains. Letting down over Disraeli Fiord, the ice shelf appeared right ahead. Its wavy surface – reminiscent of a washboard – looked as if it was yielding to the pack ice pressing against it from the north. Passing just east of Ward Hunt Island, we caught sight of another hut. All the time I was rambling on with my commentary describing what was beneath us. Flying at an altitude of 100 metres, we ran into cloud and came down to 60 metres in an attempt to get under it. I could see that instrument flying at this height in cloud was not what the pilots enjoyed. To make matters worse, we flew off the edge of the chart. I told Foreman not to worry – we were over the Arctic Ocean and there was nothing ahead for a thousand miles. He said: 'Are you sure?' I was.

Here we were only 750 kilometres from the North Pole. Peter Wadhams was delighted with the detailed profiles of giant pressure ridges that the laser altimeter was recording. Turning back towards the land, we could see that this was not the kind of weather to trifle with, so Foreman abandoned the tests and climbed to a safe altitude. As it happened, the tests were now complete and everyone seemed well satisfied with the results. We were back on the ground at Thule four hours after taking off.

Next morning the crew members, always cheerful, were more cheerful because we were going home. Thule Air Base had been generous to their guests and there were heartfelt thanks as we said farewell. As the brakes came off and we accelerated down the runway, I felt a twinge of sadness at leaving this glorious, pristine wilderness. Simply breathing the clean air had been good.

On the way home we flew over Vatnajökull in Iceland, the largest ice cap in Europe. With four companions I had tried – and failed – to cross it in 1947. We had spent weeks man-hauling a heavy sledge in the manner of the old-time explorers. All the time we were cold, wet and miserable. Now we crossed it in six minutes.

I have never been averse to progress.

9. GRAHAM LAND (1971–72)

An uncontrollable spiral dive

Our first Antarctic season of airborne radio-echo sounding in 1966–67 had proved so interesting that I decided to launch another. In 1971 I invited Dr Michael Walford, Lecturer in Physics at the University of Bristol, to handle the instruments in place of David Petrie, who had by then moved to a new job in the Falkland Islands. Mike, a former Ph.D. student at the Scott Polar Research Institute, was the pioneer of the British Antarctic Survey's radio-echo programme. He had mounted Stan Evans' prototype on a tractor and tested it at Halley Bay in 1963.[1] He had earlier been a radar technician in the Royal Air Force, and was clever at designing and maintaining electronic instruments. However, with an unkempt beard and well-worn clothes, he looked more like a tramp than an academic.

We left home on 11 November 1971 and flew via Casablanca and Recife to Rio de Janeiro. Headwinds and a diversion because of rocket firing in the South Atlantic caused us to miss the once-daily connection to Montevideo. Stranded at the same time in the Hotel Gloria was a girl in her teens who appeared distinctly ill at ease, so we decided to look after her without appearing to do so. She was Kate Hennessy, daughter of James Hennessy, the British Chargé d'Affaires in Montevideo. He had been in Montevideo since the British Ambassador, Sir Geoffrey Jackson, had been kidnapped by terrorists ten months before.

Mr and Mrs Hennessy had driven to Montevideo airport to meet their daughter, only to discover that she was not on the plane. Sensitized to the possibility of kidnapping, they began enquiries by telephoning the British Embassy in Rio. The airline reported that the girl had been seen leaving the airport in the company of two men, one of whom resembled Che Guevara. Now thoroughly alarmed, the Hennessys charged the embassy with finding the fate of their daughter. The embassy found her in commendably short order and, together with Kate, we dined the same night with two embassy officials.

On arriving next day at Montevideo, Mr and Mrs Hennessy were relieved to see their smiling daughter. As a mark of gratitude for looking after her, Mike and I were given a lift into town with the embassy bodyguard. The man cautioned us to take great care not to bump into the switch of his limousine's ear-piercing alarm system.

We boarded RRS *John Biscoe* and headed for the Falkland Islands. After the customary visit to Government House in Stanley and meetings with friends, we sailed for the Antarctic on 22 November. Our arrival in the South Orkney Islands four days later was heralded by the sight of a magnificent fleet of tabular icebergs off the western end of Coronation Island. Nearby, at Signy Island, Mike and I transferred to RRS *Bransfield*. The Master was Tom Woodfield, in whose smaller ship I had voyaged to Antarctica five years earlier. Chris Elliott was Chief Officer.

Bransfield was BAS's new, purpose-built, ice-strengthened passenger/cargo/research vessel. Though not an icebreaker, her displacement was much the same as that of the American 'Wind' class icebreakers. A diesel-electric system developing 5,000 shaft horsepower made *Bransfield* the most powerful vessel ever used by Britain in the Antarctic. Three times the size of *John Biscoe*, she was also more comfortable. A spacious flight deck aft was good for visiting helicopters – though BAS had none of its own.

Two days' steaming brought us to the American Palmer station on Anvers Island. This was not the same Palmer that David Petrie and I had visited in 1967, a stone's throw from the abandoned British hut. It was a lavish new station on Bonaparte Point on the opposite side of Arthur Harbour, 1 kilometre from the old. Two young Ph.D. students collecting insects were at the station supported by nine US Navy men. Although an unspeakable extravagance by our standards, I knew from earlier experiences at US Antarctic stations that it was their way of doing things. We put ashore three men to paint and renovate the British hut and to mark a route through crevasses to the local airstrip – the same strip that Bob Vere had landed on in February 1967.

By 1971 BAS had given up the practice of wintering aircraft at Deception Island and had taken to flying them back and forth each season between Toronto, where they were built, and Antarctica. From Punta Arenas, Chile, to Adelaide was a distance of more than 1,700 kilometres and there was always some anxiety that the pilots might encounter bad weather en route. Moreover, throughout most of the flight they would be out of sight of land and dependent on dead-reckoning to determine their position. To reduce the risk of going astray, we normally tried to station a ship in Drake Passage to act as a radio beacon. This time we were lucky in that USCGC *Staten Island* happened to be there on passage to Palmer station and she agreed to act as guard ship. Now *Bransfield*, hearing that the weather forecast for the route was good, steamed west towards Hugo Island to act as an additional radio beacon, albeit 200 kilometres off track.

BAS no longer had the Otter aircraft that David Petrie and I had used in 1967; it had been retired after its fatigue failure. John Ayers' Porter had crashed in Palmer Land in 1968. Although nobody was hurt in the crash, John had to live through an unplanned winter at Fossil Bluff. Staff at headquarters in London telephoned his wife to explain that he would be late home. 'How late?' said the unsuspecting wife. History does not record her words on hearing the answer: 'About twelve months.'

The air armada of 1971 consisted of one de Havilland DHC-6 Twin Otter and one de Havilland Turbo-Beaver. The latter was single-engined and had no de-icing system, so had to avoid clouds. They left Punta Arenas at 1130 on 29 November and at one point during the flight were forced down to only 30 metres above the briny ocean to stay below the icing level. Both aircraft flew over us and landed at Adelaide on time.

After a visit to the BAS station in the Argentine Islands, *Bransfield* reached Adelaide the following day. We could see no activity ashore. The first visitors to some Antarctic stations each spring are made to feel like unwelcome intruders because they upset the close-knit commune that has developed during the winter. The irony of the situation – which everyone knows – is that the station is totally dependent for its survival on the 'intruders' and what they bring with them. Although we newcomers were greeted in a civil manner, nobody from the station helped with unloading the ship, although she had brought all the station supplies for the following 12 months. After a day or two, however, the cargo began to disappear from the jetty. The men were doing their job – but at their own lethargic pace. Meanwhile the ship's crew worked like Trojans and put 100 tonnes ashore in two days.

One evening the Chilean naval tug *Yelcho* came alongside carrying the Commodore of their Antarctic fleet. The other half of the fleet was a slightly larger naval auxiliary – *Piloto Pardo*. *Bransfield*'s wardroom officers were royally entertained in *Yelcho*'s wardroom, after which they continued on their way. The British and Chilean Antarctic ships maintain the most cordial relations. I was told by one Chilean officer that the Royal Navy played a significant part in establishing the Chilean Navy. In contrast to some young navies we have helped, the Chileans freely acknowledge the fact.

A week passed before we were able to start flying. Tonnes of food for field parties and more than 250 drums of avtur had to be dragged up to the airstrip. BAS had generally employed seconded RAF pilots but now we had a civilian pilot, 33-year-old David Rowley. He had been a marine engineer before stopping to think one day that there must be a better way to get from A to B. The other pilot was Flight Lieutenant Bert Conchie, RAF. Rowley generally flew the Twin Otter and Conchie the Turbo-Beaver. Both aircraft used the same Pratt & Whitney PT6A-27 engine – a distinct advantage in terms of spare parts.

Several sledging parties with one or two dog teams were at work on the mainland. They had left the main sledging base at Stonington Island weeks earlier and by now were clamouring by radio to be resupplied. Also Fossil Bluff, now a wintering station for glaciological research, had to be stocked up for the whole year. At the same time I was anxious to establish a new mini-station on Spartan Glacier near Fossil Bluff, 400 kilometres south on Alexander Island. Andy Wager and Ian Rose were to make year-round observations of the heat, ice and water balance of the glacier as part of Britain's contribution to the 'International Hydrological Decade'. Glaciologist Martin Pearson and Jim Whitworth, a GA, were to winter at the Fossil Bluff hut.

Feeling the pressure, everyone was keen to get going, and on 10 December we did. In the course of a long day, the Twin Otter carried five one-tonne loads to Fossil Bluff. Each round trip took about four hours. On one of the flights we were able to make the first-ever landing on Spartan Glacier. I had discussed this project with the pilots for a couple of years but they were none too keen because the glacier flowed in a box canyon, meaning that once committed to landing they could not have second thoughts. To cope with its up-sloping surface, Dave Rowley approached straight and level before flaring out to touch down at the climb angle of the glacier. We selected a site for the hut, unloaded the cargo and took off downhill. With the idea proven, pilots happily flew in supplies to this tiny outpost for years afterwards.

Everyone at Adelaide had turned to in order to keep the aircraft supplied. The Beaver flew to Stonington and from there carried three half-tonne loads to a depot near Mount Charity, 220 kilometres farther south. By the end of the day, both pilots had been on the go for 21 hours. They were tired but happy. There followed three more marathon flying days, at the end of which most depots were fully stocked and we had carried several tonnes of rock specimens back to base. Now only some Fossil Bluff cargo remained to be taken. I sent the following to Sir Vivian Fuchs in London:

> All hands at Adelaide, Stonington, Fossil Bluff and in the field cheerfully maintained massive ground support operation to keep aircraft flying round the clock to take advantage of limited flying weather. Teamwork has been beyond praise and in the highest traditions of the Survey.

The pilots generally preferred to fly with someone in the co-pilot's seat to provide company and to help with cargo handling. In my experience most are also keen to share the flying by letting their passengers have a go at the controls. Thus initiated, several BAS employees, on leaving the Antarctic, have taken up flying as a career. I already had a pilot's licence, so was invited to do more flying than most. One day, bored by long hours at 3,000 metres over the sea, I asked Dave Rowley what would happen if an engine stopped. He replied: 'Well, why don't you just try it and find out? Pull one of the throttles, feather the blades and cut off the fuel!' Hesitating just long enough to see that he meant it, I did. The starboard propeller slowed until it was gently windmilling in the slipstream. After a few seconds, the aircraft began to veer to the right, so I compensated with left rudder. It went on turning, so I applied full left rudder. We continued to turn. Now the nose dropped and the machine rotated like a circus roundabout. In a studiously calm voice, Dave said: 'You are now in an uncontrollable spiral dive.' He was all too right. After a few seconds of feeling uncommonly humble, my plaintive voice said. 'Yes.'

Schooled in recovering from situations like this, Dave had us on an even keel in seconds. Firing up the starboard engine, he climbed back to recover the altitude that I had lost. Somewhat shaken, I did not fly any more that day. By way of explanation, he said: 'Now you know why single-engine pilots need training to fly multi-engine.' He also volunteered that it was only because the plane was completely empty and light on fuel that he had let me try it. Looking in the mirror a couple of hours later, I concluded that humility was not a bad starting point for a would-be pilot.

Radio-echo ice-sounding began on 16 December when Mike installed the system in the Otter. As the instrument itself weighed only 100 kilos, we combined cargo and science flying until every last tonne had been borne to its destination. One of the less popular cargoes was coal for the stove at Fossil Bluff. The dust permeated instruments and clothes and adhered to my balding head.

Long hours of radio-echo flying filled in more blanks on the map but we could see that it would take several hundred hours in the air to obtain a reasonable framework of known ice depths.

I carried 93 maps in a briefcase in the cramped cockpit and had to be ready to find and unfurl the next one without either blocking the pilot's line of sight or inadvertently jostling the controls. Rowley's job was to fly, while mine was to navigate the aircraft as accurately as possible. If we ran into bad weather, I had to suggest a modified flight plan and then calculate an estimated time of arrival (ETA) at the next turning point. It was fun but tiring.

One of our first flights was over Wordie Ice Shelf. I noticed that a large chunk of ice had disappeared since I was there in 1967. Although it is normal for icebergs to calve at intervals from any ice front, the scale of the change was surprising. It coincided with an area where David Petrie and I had failed to obtain bottom echoes. We speculated that brine may have penetrated the ice through rifts, making it opaque to radio waves.[2]

Another glaciological project involved measuring ice temperatures in many places. It had long been known that the ice temperature at 10 metres' depth in snow closely approximates the mean annual air temperature on the surface. We wanted to obtain a network of spot temperatures to see how the climate varied with latitude and also from west to east across the peninsula. It would be infinitely cheaper than establishing a network of manned or even unmanned weather stations.

The first ice-temperature flight was in the Beaver with Bert Conchie and Ian Rose. We landed in the middle of nowhere 15 kilometres west of Mount

Hope in the Eternity Range. After three and a half hours of laborious drilling with a hand-powered ice-auger, we found that in this place the mean annual temperature was −19°C.[3] The next flight was to the easternmost side of Larsen Ice Shelf on the Atlantic side of the peninsula. Here we were amazed to find concentrated brine in the drill hole – evidently the sea had soaked into the snow.

On the way home we approached an ice rise that I had seen in 1967. An ice rise consists of a smooth dome surrounded by floating ice. It is believed that an ice rise forms where an ice shelf runs aground on a shoal, though the shoal itself may be well below sea level. Snow accumulation progressively raises the surface until it takes the shape, though not the scale, of a continental ice sheet. At this stage it flows radially from the centre and maintains a stable profile. Ours was oval in shape, 9 by 14 kilometres and nearly 300 metres high. The seaward side appeared to rise out of the Weddell Sea pack ice. I asked Bert Conchie whether he could land on the summit so that I could take angles to peaks on the mainland in order to fix its position. He conceded that landing on a dome should be all right if he touched down exactly on the summit. So after circling to study the lie of the land, he did it with panache. I measured the angles and that particular ice rise is now on the map.[4]

Everything that Bert did was done with panache. He had been a flying instructor in the RAF and believed in precision flying. The Red Arrows needed it, but here in the Antarctic Bert still flew within 10 metres of his set altitude. It was a matter of pride. On retiring in 1975 he was the longest serving of any BAS pilot. In 1977 a glacier flowing into George VI Sound was named after him.

On 8 January 1972, a week after the ice rise landing, HMS *Endurance* arrived off Adelaide. Four years earlier she had been converted for naval use from the Danish ice-strengthened freighter *Anita Dan*. Now she was Britain's sole guardian of Empire in the South Atlantic. *Endurance* was 93 metres long and displaced 3,700 tonnes. She was immensely useful to BAS in that she carried out hydrographic surveys and did more charting in the poorly charted waters of the Antarctic Peninsula than all other ships put together. She carried helicopters, so was able to take surveyors to mountain peaks inaccessible by other means. Mike and I were invited to dinner on board as guests of Captain Rodney Bowden.

Our tractor mechanic, M.D. Macrae, had been hit in the right eye by a fragment from a shattered hacksaw blade. The station doctor, Mike Holmes, decided that he needed hospital treatment, so on 11 January we flew him to Palmer station to connect with *John Biscoe* northbound. As usual, Mike and I took advantage of the situation to operate the radio-echo sounder en route because it did nothing to slow the evacuation. In reporting the weather, Palmer had not mentioned that their airstrip was in whiteout, which is inherently dangerous because pilots cannot judge their height over the snow.

However, Rowley was by now experienced and managed to put us down gently. USCGC *Southwind*, my old friend from 12 years before at McMurdo (when she was USS *Atka*), was in Arthur Harbour and very kindly sent a helicopter to take our casualty to the station. We left him in good hands.

Palmer had soft and sticky snow, and on our first attempt at take-off we failed to reach flying speed. We got off only by trying again and following exactly in the aircraft's own ski tracks.

Faraday station in the Argentine Islands

Eureka Glacier

An ocean of ice between outcrops. The Eternity Range is on the horizon

RRS *Bransfield* off the Graham Coast

Alexander Island from Lully Foothills

Midnight in Palmer Land

A low sun put our shadow on the walls of giant crevasses

Gipps Ice Rice surrounded by Larsen Ice Shelf. It is 18 km long, 9 km wide, and 300 m high

10.   OUR PROPELLER STOPS (1971-72 continued)

With a clunk

On our radio-echo flights, particularly north of Adelaide, we noticed that in some areas there were no reflections from the bedrock under the ice. It was known that the colder the ice, the better the bottom reflections should be. Our problem was that we were trying to penetrate 'warm' ice, probably at temperatures between 0° and –5°C.

On taking off from Adelaide one day, Mike saw a strong bottom reflection on his oscilloscope but watched it fade as we climbed away from the surface. We realized that by flying low we ought to achieve much better results.

There are hazards to low flying and we were in no position to require the pilot to do it. An exercise in tact was needed because Rowley was wholly and solely responsible for our safety. So we simply explained the facts – that low flying produced better results. We did not define *low* in terms of height above the ground. After trials at various heights we were delighted to discover that the lower he flew, the safer he felt. It was a matter of perspective. At 30 metres or more he could not always perceive changes in topography ahead and could find the surface rising unexpectedly. With a lower perspective he could see what lay ahead and react in good time. The same is true for every pilot landing his aircraft: he knows that the lower he gets, the more precise his perception of height. There would be many a mangled airliner if it were not so.

However, low flying in whiteout conditions can be reckless to the point of suicide, so whenever we approached a layer of cloud or even partial overcast, we would climb to a height of at least 600 metres. There was one other condition that made us climb. Heading right into the sun, we were so dazzled by reflection from the snow that our perception of height was impaired.

Dave said that if he did make an error of judgment and the aircraft's skis touched the snow, he believed that the machine would just bounce back into the air. There is a place in the Antarctic called Touchdown Hills, so we would not be the first.[1] However, in several years and hundreds of hours of low flying since these first experiments in 1972, we have yet to brush the surface.

On 13 January we acquired an excellent ice-depth profile up the 80-kilometre-long Fleming Glacier, crossing some chaotically crevassed areas where we had failed to obtain any results while flying at 300 metres. Combining operations as usual, we then took supplies to a sledging party on

Clarke Glacier. Bob Wyeth and Brian Hill had unwittingly camped beside a large snow-bridged crevasse and asked us to taxi across to them. We invited them to come to us instead. Their supplies were down to four days of food and fuel. Considering that weather sometimes prevents flying for a week or more, I was unhappy that they had not asked us to come earlier. Keeping an ample reserve of food is a very basic safety measure – particularly in the Antarctic.

At this stage of the season, three well-known glaciologists working on Livingston Island in the South Shetland Islands had been imploring me to determine the thickness of the ice cap there. They wanted to obtain ice cores to study climatic change. I had been involved with the planning of their project and was equally keen to know the answers. Moreover, all were friends and had helped me in the past. As often happens in our business, theirs was an international team. Valter Schytt, from Sweden, had been my boss in Antarctica 20 years before; Colin Bull, from Ohio, had wintered in Greenland; and Olav Orheim, from Norway, was a Ph.D. student at the Ohio State University.

One day they sent a radio message reporting that their weather was good for flying. But with the limited range of the Otter, we knew it would not be easy to reach them. The round trip distance from Adelaide to Livingston Island was 1,400 kilometres, which was much greater than our normal operating range. So we filled every fuel tank and then loaded six drums into the cabin, taking a hand pump to transfer the cabin fuel to the main tanks. Our intention was to land on Livingston Island so that one of the party could guide us on the sounding runs. I had asked them to mark a runway. However, when we arrived over the island there was a full whiteout and it would have been suicide to attempt a landing.

On the way home we had to climb to 4,100 metres to stay clear of cloud. Sitting quietly in the cabin at this altitude was no problem, but vigorously pumping fuel without supplemental oxygen had each of us gasping for breath whenever we took a turn. The weather improved as we flew further south, making it possible to do several hundred kilometres of low-level sounding, so the flight was not without value. However, it was late at night by the time four very tired aviators stepped onto the ground. We decided, there and then, that the risks inherent in marathon flights like this could not be justified.

On particularly windy days when we were low flying – 10 metres over the ground – we noticed a mild high-frequency vibration in the aircraft. For some days we pondered the reason before concluding that it came from the influence of sastrugi on the 'ground-effect' – the compression of air between the snow and the wings in flight. It was easy to demonstrate this by climbing – the vibration invariably disappeared.

One windy day Dave Rowley made a tactical error by flying close on the downwind side of a sharp rock ridge. The Otter was severely shaken by a vortex coming off the summit. Two heavy car batteries jumped off the floor and came down again with a crash. The stall warning sounded although we were flying at normal cruising speed. We ourselves were strapped to our seats, so apart from moments of levitation, only our nerves were frayed. The lesson, we concluded, was: 'Thou shalt not fly downwind of a mountain ridge in windy weather.' We never did again.

On 22 January we went camping. Professor John Nye, one of Mike Walford's colleagues in Bristol, had suggested that the radio-echo sounder might be used to determine the speed of movement of a glacier. The idea needed testing. It could be immensely important as a means of measuring the rate of flow on remote parts of the ice sheet where there are no fixed points for reference. I chose Fleming Glacier for the experiment on the grounds that it was deep and should be fairly fast-flowing. While Mike wielded his instrument, I could independently determine the rate of movement by taking theodolite angles to nearby nunataks – there were plenty within a few kilometres.

Dave Rowley flew us to a lovely smooth landing in the middle of the glacier where – we hoped – there were no crevasses. Paul Burton came along as GA. Mike and I got to work while Paul pitched two tents. On finding that the glacier was 1,090 metres deep, Mike noticed that moving the radar antenna only a few centimetres altered the shape of the returned echo. The pattern of echoes gave a framework against which glacier motion might be detected. Of course the glacier was moving and our camp with it. After a few days Mike moved the sounder upstream a short distance until the pattern was the same as when he first observed it. We thought that the horizontal distance on the surface between the two sites would give the rate of ice movement.

It was an exciting moment when, after four days, we compared my calculated rate of movement with Mike's. My measurements gave 46 cm/day whereas Mike's gave 38 cm/day. Neither of us laid much claim to accuracy, so we concluded that the figures agreed within the limits expected from such a short-period experiment.[2]

While waiting for the glacier to move far enough to measure, I had drilled into the ice to a depth of 10 metres to measure its density and temperature at each metre on the way down. The 10-metre temperature here was –12°C.

Returning to Adelaide, we spent the next two weeks busily resupplying field parties, carrying rocks for geologists, ice-depth sounding, and landing in all sorts of places to measure ice temperatures.

It was the warmest time of year and the inhabitants of Fossil Bluff were sometimes seen topless on calm days, basking in the sun. At this season it was not unusual for the thermometer to read a degree or two above freezing. Most of the time it stayed between 0° and –5°C in daytime, falling to –10°C on clear nights.

Glaciologists do not like hot weather. Too many days above zero could affect the substance of their living. The eighth of February was a particularly warm day that brought a series of potential disasters. After a successful four-hour sounding flight in the morning, we landed at Fossil Bluff to refuel. That done, Dave was taxiing the Otter towards the skiway when its left main ski broke through the snow into a sub-surface pool of melt water. The wing-tip dropped nearly to the snow surface. We could see that if, at any stage of trying to recover the situation, half the weight of the aircraft came to bear on the wing-tip, the machine might never again be airworthy.

If that kind of accident happened at home, it would be a straightforward job to rebuild the wing or to attach a new one. But with only the Beaver to back us up, there was no way to rebuild a wing on the spot nor to deliver a new one. After a lot of digging, we dragged the aircraft with a 'Muskeg' tractor back onto a firm surface, then breathed a sigh of relief that BAS had been spared the million pounds or more that it would have cost to replace the machine. It had taken two hours of hard manual labour to get back to where we had started.

The second flight of the day lasted three hours and took us to the western extremity of Alexander Island. Although I was using the most recent maps to navigate the aircraft, we were often reminded that the age of exploration was not over – as many people claimed it was. Flying the length of Latady Island, an oblong ice rise shown on the map as about 60 kilometres long, I could not reconcile my dead-reckoning calculations with what we perceived through the windscreen. Either I was befuddled by exhaustion or the map was very wrong.

If I had made no mistake, the island must be more than 100 kilometres long. But could this be? In reading accounts of aerial exploration of the Antarctic I had become sceptical of the claims of aviators to new discoveries. Aircraft can be pushed all over the place by winds – we knew our speed through the air but had no way of determining our speed over the ground. I doubted my own arithmetic, but on referring to recent weather satellite images of the area I found that Latady Island is 110 kilometres long. We afterwards published a new map of Alexander Island showing this and many other changes.[3]

Landing back at Fossil Bluff, we had been airborne for seven hours since taking off in the morning. Most of the time we had been flying at low level, knowing that a moment's inattention could be our last. We were tired and ready for bed. But as in my dog-sledging days 20 years before, we knew that the first thing to be done on making camp is to feed the beasts of burden. With dogs it is because they are ravenous; with aircraft it is to avoid the risk of overnight condensation in the fuel tanks. So we refuelled and then jumped aboard to taxi the aircraft to its tie-down anchors.

Dave pressed the starter for the left engine. With free-turbine engines there is always a delay before the airflow starts to rotate the propeller. But this time, as the turbine speeded up, the propeller stood stock still. There was something *very* wrong. Dave shut down the engine and jumped out to see if the propeller could be moved by hand. Normally it can be swung with the little finger, but today there was an ominous 'clunk' as if a piece of metal had jammed in the gear wheels.

Trained as a marine engineer, Dave quickly realized that no sharp tap with a hammer could solve this problem. We were grounded, finally brought down to earth. It was now late in the season and Bert Conchie's Beaver was the only thing standing between us and an unwelcome ten-month sojourn at Fossil Bluff. The three winters I had already spent in the Antarctic might now become four. We could only pray that the Beaver, with its single engine – identical with our remaining engine – would not fail.

There was no spare engine within 1,000 kilometres. A new engine was on its way south in RRS *Bransfield* but it was not due to arrive at Adelaide for another couple of weeks. Dave Rowley and the aircraft engineer, Dave Brown, built sheerlegs over the dead engine with a tarpaulin to protect them from wind while dismantling the engine. After a day's work three small pieces of chewed metal were found in the gearbox. Now we knew. If those pieces had found their way between the gear wheels in flight, the engine could have exploded. It was better not to think about it. I remember once being warned not to linger in 'the turbine disintegration zone' of a military aircraft. Mike Walford had just spent a good many hours sitting in it.

Although it would have been possible, Brown deemed it too risky to reassemble and use the engine until the manufacturers could dismantle it at their factory and explain what had gone wrong. So I asked the Beaver to come to Fossil Bluff to remove people who, in the circumstances, were useless – meaning me and Mike Walford. Bert Conchie flew in with the second engineer, Rob Campbell-Lent, to help with the engine change, and Dave Rowley flew us back to Adelaide. With hindsight, it might have been safer to mollycoddle the Beaver until it could bring in what mattered most – the new engine.

John Biscoe was at Adelaide but now steamed north to fetch the engine from *Bransfield* in the South Shetland Islands. This excursion took four days – she returned on 24 February. The engine was in a container far too big to fit in the Beaver and carefully suspended so that its components would not be damaged in transit. The normal procedure would have been to lift it into position straight from the container. Now we had to hand-carry it to the Beaver, angle it through the side door and delicately lay it on a bed of sleeping bags. No baby, I thought, was ever handled so gently.

Dave Rowley took it to Fossil Bluff, where it was hoisted into the empty nacelle. It took a week to install and test-fly. There were wet eyes – from relief and joy – when finally our beautiful Twin Otter buzzed *Bransfield* just as the ship was arriving at Adelaide to carry us home.

The Governor of the Falkland Islands, Gordon Lewis, together with his wife, had come in the ship. At that time it was very rare for a woman to be seen in the Antarctic. Welcome as she was, such visits caused consternation on the base because of the total lack of civilized bathroom facilities. I think she was briefed not to ask. Men were accustomed to relieving themselves in the open just outside the front door. Other visitors arriving at the same time were Ray Adie, Roy Piggott, Bill Sloman and Paul Whiteman, all senior officers of BAS who were enjoying a grand tour of the bases.

Both Stonington and Adelaide were equipped with dog teams. In the field, they were fed dehydrated rations, but on base they lived on seal meat. Having very little shoreline that was not covered by ice, Adelaide was a bad place for hunting. *John Biscoe* generally took time for an annual mass slaughter in the fjord region to the east. Now we had 134 seal carcasses putrefying in a stream of melt water, so on warm days the stench was nauseating.

Meanwhile, *Endurance* was surveying uncharted waters, as she was required to do, and had a nasty scare off Red Rock Ridge on the mainland. The ship ran hard and fast onto an uncharted pinnacle of rock. Although she was holed, it was only through the outer plating of her double skin. The ship came off on the next tide – a merciful delivery. Antarctica is not a good place to be shipwrecked.

Mike Walford and I moved aboard *Bransfield* on 5 March. Our flying season had had its ups and downs but we had no complaint. As usual, we could not have done it without the help of many people. We had added some 7,000 kilometres of ice-depth soundings to the map and believed that we had found a new way to measure the speed of glaciers.[4] At the request of the BAS Geology Section in the University of Birmingham, we had sounded about 1,500 kilometres along traverses covered in the course of their gravity and magnetometer surveys. At the request of John Behrendt of the US Geological Survey, we had sounded 167 kilometres along the track of the US Antarctic Peninsula traverse of 1961–62. In addition, we had done sounding runs along the ice-deformation survey tracks of glaciologists wintering at Fossil Bluff. One of the more exciting runs was down Spartan Glacier from its headwall, involving an extremely steep descent.

Five days later, on dry land at Punta Arenas, Chile, we feasted in restaurants and went sightseeing. Just down the coast is a place where the remains of sixteenth-century Spanish colonists have been found. A nineteenth-century fort has been carefully restored as a tourist attraction. Right on the shores of the Strait of Magellan there is a tiny graveyard where sailors who died on the 1826–30 voyage of HMS *Beagle* are buried. One wooden cross bears the words:

IN MEMORY OF COMMANDER PRINGLE STOKES R.N.

HMS BEAGLE

who died from the effects of the anxieties and hardships incurred

while surveying the western shores of Tierra del Fuego

Aug. 8th, 1828

On the northern outskirts of Punta Arenas we visited Instituto de la Patagonia, a cross between a research institute, natural history museum, folk museum and transport museum. They have a small zoo with a good collection of native fauna.

Flying north, I visited several institutions with Antarctic interests. First there was Instituto Antártico Argentino in Buenos Aires. Strangely, from our point of view, the institute is part of the Ministry of Defence. The Director, General J.E. Léal, introduced me to a number of their scientists.

I felt most at home with René Dalinger, Léal's Chief Scientist, who was a glaciologist and an old friend. René was born in Peru of missionary parents, read geology at Universidad Nacional de Córdoba, and worked professionally as a hydrologist. This led him to glaciology because some of the water used in the Province of Córdoba originates from glaciers in the Andes. When the International Geophysical Year of 1957–58 (IGY) was being organized, Dalinger was the obvious choice to winter in the Antarctic as his country's glaciologist. He spent a year at General San Martín Base in Marguerite Bay. The following year he had to retire from fieldwork after being involved in a serious helicopter accident. However, Fuerza Aérea Argentina sought his help in finding a runway site in the Antarctic for aircraft not fitted with skis. René recommended Seymour Island, which later became Vicecomodoro Marambio Base, the first hard runway for transport aircraft in the whole of Antarctica.

From Buenos Aires I flew to Washington, DC, to visit colleagues at the National Science Foundation and the Arctic Institute of North America. Another friend was Bill Macdonald of the US Geological Survey. At the time, Antarctic mapping owed more to Macdonald than to any other person. He had served as navigator on many aerial survey flights and was now preparing maps of the Antarctic Peninsula. We needed his maps to be able to plot our ice-sounding flight tracks. Although I had recorded hundreds of positions in relation to mountains and nunataks, it had become obvious that some of the features were not where the old maps put them.

Each of these visits included important exchanges of information on the work that we had been doing. I never cease to marvel at just how international in character polar research is and has to be. For most of us there is no career structure nor is there any guarantee of employment. To remain in glaciology, it had been necessary for me to work in Canada 1956–57 and in the US 1959–63 because there were no posts for glaciologists in Britain. I was 45 years old and working in Cambridge; BAS had only recently offered me a post which in due course was to become permanent. I was lucky, because most BAS employees are on short-term contracts. I have met many people who envy my kind of life but very few who – blessed as I was with a wife, three children and a mortgage – would sign on without some guarantee of tenure. To me, however, it was all worthwhile. I was fortunate to have a wife who agreed.

11. SPREADING OUR WINGS (1974–75)

Flying with a polar legend

By 1973, the science of glaciology was becoming more advanced. I felt that, without additional professional staff, the glaciologists at the SPRI could not hope to stay close to the forefront of our field. We had some excellent young men but they were on short-term contracts and BAS had no long-term posts to offer. I was lucky to have Dr David Peel, a chemist from the University of Bristol, who had joined us in 1969. In due course he was to become the top specialist in Britain on the chemistry of Antarctic snow – or more precisely, the chemistry of impurities found in snow.

Recent developments in radio-echo sounding offered more exciting possibilities, but advertising for a 'radio-echo sounder' would get us nowhere. There was no such thing. My best bet was to find a good physicist who, though without expertise in glaciology, was keen to learn. So we advertised widely. Making it clear that long periods might be needed far from home, deprived of the joys of life in England, we had few applicants. But one of them seemed just what I was looking for. Christopher Doake had earned his Ph.D. in low-temperature physics with a thesis on the motion of charged ions in liquid helium. The low-temperature bit sounded like an ideal background until he reminded us that liquid helium boils at a temperature of –270°C. In Antarctic summers we would be operating at around –10°C. From his low-temperature background, Chris had moved into computers, and when he applied to join BAS, was involved in strategic planning at Rank Xerox.

I made clear to him that although initially he was needed for radio-echo sounding, my experience was that a scientist performs best when set free to follow his own ideas. The only constraints were that it must be good science and it must be Antarctic science. There would always be envious people looking over our shoulders – hoping to convince our government masters that their own line of research had greater merit than ours. Polar research is expensive, and throughout my life I have had to do battle with those who pointed out – not without justification – that a dozen or more scientists could be employed in a British university for the cost of one in Antarctica.

BAS offered Chris Doake a post and he joined us in January 1973. Although at this stage I and my staff were still at the SPRI, BAS was planning to build its own headquarters on the west side of Cambridge.

103

I gave Chris two years to learn something about glaciology and to develop ideas for fieldwork in the 1974–75 season. But as a main task, I asked him to handle what David Petrie did in the 1966–67 season and Mike Walford did in 1970–71 – the technical part of radio-echo sounding. As usual, I would look after the navigation. He was keen to get his teeth into some field-work and so was I.

We were embarked on another season of airborne ice-sounding – but this time we must go further afield. BAS policy with respect to its aircraft had been rather like that of mother hen – there were squawks from headquarters if we ventured beyond narrow limits hallowed by tradition. The traditions had developed over a long period during which only single-engined aircraft were flying. Back in 1947, the first BAS aircraft was a three-seat Auster. For 22 years thereafter only single-engined aircraft were flown. Sometimes one aircraft was operating without back-up, as I did in 1966. It was common sense not to go too far afield.

The change came in 1968 when BAS acquired a Canadian-built de Havilland DHC-6 Twin Otter. A second Twin Otter was purchased in 1972. Now there were two aircraft with twin engines and we could be more ambitious. However, budget problems had led to one of them being hired out to the US Antarctic Program from time to time and for those periods we were back to one. Flying right across the continent to McMurdo Sound – sometimes non-stop – finally made clear that the aircraft were capable of operating much further afield than was customary. In other words, our modern and very expensive aircraft were being under-utilized.

Dave Rowley had left BAS to work for an airline. Bert Conchie was to spend part of the season flying one of the Twin Otters for the Americans on the Ross Ice Shelf. A new pilot, 26-year-old Giles Kershaw, was to fly the second Twin Otter. With only one aircraft to look after all field parties during the first half of the season, there would be no time available for radio-echo sounding until Bert Conchie flew back from McMurdo.

Chris Doake was already on his way south but I decided not to leave Cambridge until 1 January 1975. I flew via Paris, Rio de Janeiro and Buenos Aires to join RRS *Bransfield* in Montevideo. Stuart Lawrence was Master. A young Argentine scientist, Pedro Skvarča, came to spend the season with us for training in glaciology. Besides our own people, the ship was carrying 12 Americans and some cargo bound for Palmer station. We anchored in Arthur Harbour next to the station on 11 January. I went ashore and was shown over the station by Dr George Llano, the man in charge of funding all biological research at the US National Science Foundation.

Bransfield reached Adelaide on 14 January without being slowed by pack ice. The place was tidier than I had ever seen it and I wondered what could have brought this normally listless community alive. It soon became apparent. The Base Commander, Steve Wormald, ruled with an iron hand. Though fair-minded, he knew when laziness had postponed jobs that needed doing and promptly galvanized people into action. It did not make him popular but the effect had been cathartic. I concluded that he had a great future.

For some years now, the aircraft skiway had been deteriorating and now it was quite bumpy. Someone suggested that Rothera Point on the east side of Adelaide Island could be a possible site for a future hard (bare-ground) runway, so I asked *Bransfield* to take us there. The approaches to the point were uncharted, so the ship stayed well clear while five of us went ashore in an inflatable boat. The whole area consisted of raised beach cobbles and gravel, and it was remarkably level. There was only one possible alignment for a runway and we marked it out with a few small rock cairns. At one end there was a hazard from grounded icebergs but we felt that pilots could live with it. Luckily Alan Smith, the man responsible for all BAS building projects, had come ashore with us. Big Al, as he was known, now set to work with a spade to see whether the beach material had solid foundations. Together we concluded that most of the necessary levelling could be done with a bulldozer. It was certainly the best potential runway site that any of us had seen in this part of the country.[1]

Back at Adelaide, Pedro and I moved ashore. We were delighted to find that Chris Doake, who had come south some weeks ahead of us, had already installed the radio-echo sounder in one of the Twin Otters. The second aircraft, piloted by Bert Conchie, was now back at Adelaide having flown from McMurdo via the South Pole. He would handle the supply of all field parties until the end of the season.[2]

The latest version of the radio-echo sounder had been built in Cambridge by Hugh Macpherson, a young electronics engineer who had come south with us to learn the trade. We planned to fly with a science crew consisting of Doake, Macpherson, Skvarča and me. Most of our work would not involve landing away from Adelaide, so Conchie decided that the safest introduction to Antarctica for Giles Kershaw would be to fly with us. I would have been happy to work with either of them.

Giles had spent the first 13 years of his life in India, where his father managed tea and rubber plantations. After schooling in England he had applied to join the RAF but failed the aircrew medical because of his eyesight. Undeterred, he learned to fly and worked his way up to a commercial licence. When he became a company pilot, resplendent in captain's uniform, his juvenile appearance once led him to be mistaken for the doorman.

We were unlikely to make that mistake. Free from the constraints of civilization, he sported an unkempt beard and the oldest clothes he could lay his hands on. But his looks concealed a quiet, well-spoken man who was the soul of politeness even on occasions when others spoke harshly.

Giles was ready to fly and so were we. Our five-strong team was airborne in Twin Otter VP-FAP within 24 hours of my coming ashore. While I navigated in my usual, old-fashioned way, we had on board a Litton-51 inertial navigation system (INS). This would be far too expensive for BAS to acquire but it had been lent by the Americans for Conchie's work on the Ross Ice Shelf. Much as we liked to have such a gadget, I had to remember that in principle, it was no more than a dead-reckoning device whose errors accumulated with time. I still needed to record our positions as often as possible with respect to landmarks.

As I had done with Dave Rowley three years before, I delicately explained the merits of low flying for our work but left Giles to decide on the safe height above terrain. He experimented at various heights before concluding – as Rowley had done – that lower was safer than higher. He confessed to enjoying the challenge and by the end of the day was flying 10 metres off the snow. We went east to the Larsen Ice Shelf and I noted that all on board quickly began to function as a team, though Pedro had little to do but admire the scenery.

Unfortunately, a technical problem with the oscilloscope camera was found afterwards to have ruined the day's work. I was always conscious of the cost of what we were doing, so failure depressed me.

Pedro came into his own the very next day. I had arranged with his bosses in Buenos Aires to use Argentine fuel when operating in the vicinity of Argentine stations on the Weddell Sea coast. In exchange, we would do whatever ice-depth measurements they wanted. The weather report being good, we flew over the Graham Land plateau to a small station called Matienzo on Seal Nunataks. None of us knew exactly where to find it, but in the event, we found several small buildings on Larsen Nunatak. We buzzed them to rouse the occupants. There was a skiway marked by a line of fuel drums on the ice shelf 1 kilometre south of the nunatak, so Giles landed on it. A Snocat approached from the station.

Argentine stations in the Antarctic are unashamedly military outposts belonging to either the army, navy or air force – never mixed. Article I of the Antarctic Treaty, however, prohibits military bases. This little difficulty is overcome by maintaining a faint veneer of science.

Matienzo belonged to Fuerza Aérea Argentina. The inhabitants appeared surprised by our arrival, perhaps slightly suspicious, so we pushed Pedro to the front to do the talking. Pedro talked at length – he always did. But we were relieved to see scowls gradually replaced by smiles, then handshakes, after which we were invited to come to the station.

Larsen Nunatak is a low ridge 1 kilometre long, obviously of volcanic origin and looking as if it had only recently erupted through the ice shelf. The station consisted of a few small buildings and supply dumps. A cliff on the far side was conveniently situated for disposing of rubbish, which was carried seaward by the moving ice shelf. We were invited into the living quarters and given coffee with a shot of Old Smuggler whisky. Giles, I noted, declined the whisky; I found afterwards that he was almost teetotal.

There were seven men on the base, led by Ricardo García. We arranged radio frequencies for contacts in the coming weeks and asked them to switch on their radio beacon if we needed it. By the end of the visit the social ice was broken and we felt among friends. The weather had deteriorated to a 60-metre ceiling with whiteout but Giles said 'Let's go!' So we climbed through the gloom and returned to Adelaide. The radio-echo sounder seemed to function well but Pedro developed the recording film and found it blank. I was shocked that we had failed for the second time. Surely we can do better, I thought, with two highly qualified people on board to check everything? But I sensed that harsh words would not help, because all of us were embarrassed.

After a day of trying to track down the problems, we combined a trial flight with an aerial reconnaissance of ice conditions for the benefit of *Bransfield*, which hoped to steam north through the inside passage to the east of Adelaide Island.

The trial flight was a success. Now we became more ambitious. Before coming to the Antarctic I had arranged with the US National Science Foundation to let us have 20 drums of aircraft fuel at Siple station, a small outpost 1,200 kilometres south of Adelaide. Taking transport costs into account, all fuel flown from McMurdo to Siple, as this was, probably costs as much as Scotch whisky in Scotland. It was not a coincidence that the decision to help us was made by Ken Moulton and Bob Dale, old friends from my years at McMurdo (1959–62).

Siple was too far for one hop, so we would refuel at Fossil Bluff on the way south. Although we always started out with every tank full, the distance to Siple was such that before reaching it, we would be beyond the point of no return. Pilots have an understandable aversion to this when flying to an unfamiliar destination and Giles was no exception. He had arranged with the engineer Alec Simon, known as Slim, to fit two ferry tanks in the cabin. We took off with 2,280 litres – nearly two tonnes of fuel. Now, come what may, we could get home. Slim came with us in case there was any problem with the aircraft. He had packed a trunk full of aircraft spares and technical manuals. Chris Doake brought along plentiful supplies of film for the radio-echo sounder.

Siple reported fine weather on 21 January and we were off the ground at 1310. The terrain south of George VI Sound was new to me and new to BAS, but we did have sketch maps made from American aerial photographs and it was not difficult to keep track of our progress. Siple station was on moving ice 150 kilometres from the nearest nunatak, so flying just above the surface, we had to keep our eyes peeled. We found the station but saw no skiway – Giles landed anyway.

Evidently because of some Arctic experience, the station crew had laid out a green tarpaulin and now motioned to us to taxi onto it. I had known ski aircraft to taxi onto boards to prevent their skis freezing to the snow, but

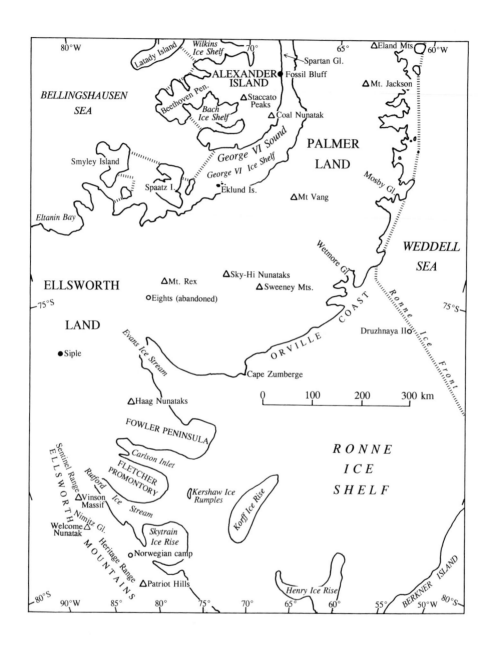

Map 7   Ellsworth Land and Ronne Ice Shelf

108

never tarpaulins. Giles was mystified, but being new to the Antarctic, thought that it must be the thing to do.

Coming to a full stop, I noticed that the radio altimeter said we were 65 feet (20 metres) off the ground. If we had relied on it, we would have crashed some hours before, but we were aware of the history of the problem. A false reading like this – my diary noted – 'has killed many an airman'.

We were greeted by a very friendly bunch of Americans, led by Jeff Harbers, the Station Manager, and Dale Merrick, Station Scientific Leader. It was some time since they had seen any new faces, so our arrival meant a welcome excuse for a party.

However, my mind was on other things. Before landing, I had sighted the Ellsworth Mountains 180 kilometres away. The weather was fine and the forecast was fine. Siple had a reputation for miserable weather; so while we were tempted to accept their hospitality, I said we would stay only for dinner. We had a preference for 'night' flying when the sun was low in the sky because long shadows made it much easier to judge the nature of the terrain ahead.

The station was well below ground level. We were ushered down two slippery snow staircases and into a brightly lit building. The huts had been built on the surface but had been progressively buried by snow accumulation. By BAS standards the accommodation was luxurious. There were squeals of delight when one of our party discovered flushing lavatories – something altogether unheard of at Adelaide. We marvelled at the extravagance of melting snow, only to flush it down a drain. Like the broken egg-yolk at Thule Air Base, it was a reminder that here there were different priorities.

Depriving Siple of their late-night party – or at least postponing it – we refuelled the aircraft and left Slim at the station to get some sleep. My ambition now was to fly right round the Ellsworth Mountains while continuously recording ice depths.

For years I had longed to see this area. The range was first sighted on 23 November 1935 in the course of the first trans-continental flight. Herbert Hollick-Kenyon, a Canadian pilot, with Lincoln Ellsworth, scion of a wealthy American family, flew a single-engined Northrop monoplane from an island off the tip of the Antarctic Peninsula to R.E. Byrd's old wintering station, Little America on the Ross Ice Shelf. They were the first to glimpse these mountains. Four landings were made in the course of their crossing. It was a courageous feat because if they had failed, there was nobody in a position to rescue them.

Jeff Harbers asked if he could come with us for the ride. Considering what his crew were doing for us, I felt we should do our best. I well knew the monotony of spending months at a time with nothing but featureless snow on all sides.

After starting the engines we had to wait 15 minutes for the INS to align with the earth's axis. As soon as it had, Giles looked back into the cabin and asked 'All set?' Four heads nodded. Giles opened the throttles to taxi but the aircraft stayed put. He upped the stakes to take-off power and heaved back and forth on the stick, but still nothing happened. We were not frozen to the tarpaulin, because there was no snow on it. There was just too much friction. Giles glanced at me with a look of despair, and shut down the engines. With spades and ice axes, we dug under the skis and tugged at the green canvas. Finally we tore the tarpaulin to shreds and freed it piece by piece. We concluded, and Jeff Harbers concluded, that we live and learn – at a price.

We took off at 2344 and headed south at 120 knots. This was the most economical airspeed for the Twin Otter and the safest for low flying. After a while the great Sentinel Range loomed ahead. I had good maps of the mountains that the Americans had recently published. Towering above us were ridges interspersed with broad valley glaciers flowing eastwards. Behind them rose Vinson Massif, the highest point in Antarctica (4,897 metres). We were flying over what appeared to be an ice stream paralleling the mountain range.[3]

We all knew that just being here was something special – a privilege few men could have dreamed of. The radio-echo sounder was functioning well, recording ice depths of more than 1,000 metres in places.

Beyond the Sentinel Range there was a gap in the mountains from which flowed a big glacier named for Admiral Chester W. Nimitz of the US Navy. Racing along at our usual height, we were surprised to come across the tracks of a snowmobile.[4] So we were not alone in this wilderness. I said: 'Let's follow them!'

I guessed whose tracks they were. As a precaution against any possible emergency, my usual practice was to find out ahead of time whether anyone was in the vicinity of our proposed flights. I had established that apart from Siple, there was only one field party within 1,000 kilometres. This comprised a Norwegian geological expedition flown here in a US Hercules from McMurdo – a classic example of the sort of international cooperation that also allowed us to be here. Coming over a low rise, we encountered four green pyramid tents. Evidently they were not expecting visitors from the sky or anywhere else. A partially clothed man emerged from a tent, so Giles landed beside the camp.

Transported with excitement at this encounter, we had overlooked the fact that normal people use the night for sleeping. It was two hours after midnight. With appalling suddenness, we had wakened them from a sound sleep. Three more mystified faces peered out of tents and then emerged, dressing. Recognizing one of them, I realized what a tight little band we Antarcticans are. The geologists were Thore Winsnes, Audun Hjelle, Yoshihide Ohta and Kjell Repp. I had met the first two at Norsk Polarinstitutt in Oslo. The group had been flown in with four tonnes of supplies and were now waiting to be

Pilot Giles Kershaw

The Gullet, a narrow channel separating the mainland (left) from Adelaide Island

Land access to the interior is often blocked by ice cliffs, crevasses, and mountains. This is Mount Paris, Alexander Island

Christopher Doake

The author

We learned to hold the shadow in the right spot – not too near and not too far. Here my fast shutter-speed appears to have stopped the propellers

We found two small nunataks in a blank on the map. The aircraft's ski tracks are above the lower summit. Vinson Massif appears hazy on the horizon – above the ski tracks. No wonder – it is 240 km from my camera

We climbed the rock and photographed ourselves – triumphant

taken home with four tonnes of rock specimens. They had dug a snow cave to keep frozen food from thawing in the sunshine. Fresh bread and fresh meat were in plentiful supply, together with a great variety of other foods. With an elaborate radio aerial they had talked not only with McMurdo but also with contacts in Europe.

I was scarcely aware that among the ten of us gathered round the tent were citizens of Argentina, Japan, Norway, the United Kingdom and the United States; and the aircraft was made in Canada. In this corner of the planet our origins were irrelevant. The Norwegians wanted to show their hospitality but I wanted the minimum stay consistent with good manners. They offered us coffee and sherry. Apologizing for disturbing them, we jumped back in our machine and fired up the engines. The inertial navigation system refused to align with the earth's axis – evidently we were too close to the South Pole. This was worrying because our magnetic steering compass had also packed up. Giles was left with the standby compass, the same size as some of us use in a car back home. However, it was plain that the only thing that could worry him was an engine failure – and even if that happened he would nurse the machine home on the other one.

After 70 minutes on the ground we were again heading south, skimming over undulating ground at the foot of the Heritage Range. As usual, we cut as close as possible to nunataks in order to position our track on the map afterwards. Soon we turned right to head for the furthest south outliers of the 400-kilometre-long chain of the Ellsworth Mountains.

A few minutes later both Giles and I noticed to our right a large area of smooth snow-free ice, its vivid blue colour contrasting with the all-pervading whiteness of the landscape. It was in the lee of a ridge called Patriot Hills. Neither of us spoke but the scene became etched in our minds. Thirteen years later this patch of ice was to witness the start of a far-reaching development in Antarctic aviation – in which we ourselves would play a part.

Now we were at our farthest south, $80\frac{1}{2}°$ south latitude. Heading up the west side of the range, we had to climb to follow rising terrain. Opposite the highest peaks of the Sentinel Range the aircraft was hugging a snow surface that was itself 2,600 metres above sea level. To our right lay a chain of peaks 4,000 to 5,000 metres above sea level – higher than any peak in Europe. Vinson Massif, the highest, was dissected by steep valley glaciers flowing between razor-sharp ridges. The mountain was named for a politician from Georgia who supported Antarctic exploration. Next came Mount Shinn, a pyramidal peak of the kind that beckons mountaineers. Lieutenant Commander Conrad S. (Gus) Shinn, USN, was pilot of the first aircraft to land at the South Pole. Then Mount Tyree, named after my friend of 1960–62 at McMurdo, Admiral David M. Tyree. The west face of Tyree was a 2,800-metre sheer wall of rock. One day it too would attract climbers, but at the time I felt that anyone trying it should be locked in a strait jacket until he saw sense.

111

By now the sun had come round to the port quarter – south-west of us – and cast a fine shadow of the Twin Otter just off my side of the aircraft. Giles asked me to take control while he had a snack. I could see the shadow out of the corner of my eye and it was the best way to judge height. If the shadow crept away, I was flying too high. If it came closer, I was descending. If the shadow's left wing-tip was lost to view beneath us, I was too low. We learned to hold it in the right place and soon developed a reflex response to any change. Throughout the season we were to find that, whenever possible, keeping an eye on the shadow was an important safety factor to guide or check on our perception of height.

The peaks to our right became progressively lower as we approached the north end of the Sentinel Range. Our departure fix was recorded as we passed 50 metres off three tiny nunataks marked 1390± metres on the map. From here there was only featureless snow the whole way to Siple, 180 kilometres away. Giles was holding the aircraft as steadily as he could but there was some turbulence being transmitted from the sastrugi below. He was too busy flying to be able to look at any of his instruments. He had taught me to read them and to report any anomalies. The standby compass, which was all we had to steer by, was swinging wildly. Keeping my eyes on the card, I noted its reading at each end of the swings. Taking half the angle between the ends gave the best estimate of our course. I told Giles 'Left a bit' or 'Right a bit' to keep us on our intended heading of 353° magnetic.

It worked. After another 45 minutes Siple appeared right on the nose. We had been airborne for nearly 9 hours since leaving Adelaide 17 hours before. I had recorded, and timed to the nearest second, about 150 events to help with later plotting of our track. Chris and Hugh had successfully measured ice thicknesses most of the way. With their eyes glued to an oscilloscope, they had been unable to share the much wider view that Giles and I had enjoyed from the cockpit.

We were tired but euphoric about everything that we had seen and done. Jeff Harbers told me later that it was the greatest day of his life – and I know he meant it. To Giles and the rest of us, it was just *one* of the greatest.

12.   DISCOVERY (1974–75 continued)

We stood on the summit – triumphant

By now we had used 17 of the 20 American fuel drums so generously offered to us. By all rights, we should be heading home, but the locals did not want us to leave. Despite waking them at all hours of the night, we had brought a sense of excitement that was evidently lacking in their daily lives. On digging up some fuel that had escaped their inventory, Harbers sought permission from McMurdo to give us an extra ten drums. McMurdo, bless them, agreed. If this was our reward for giving Harbers the flight of a life-time, it was out of all proportion – though doubly welcome.

After a meal and a good sleep, we were off again. Our habit of flying when the weather was good, instead of according to clock time, rather baffled our hosts but they remained courteous. Different clock times added to the problem. In the aircraft we kept Greenwich Mean Time. In the air again on 22 January at 1846 GMT, it was 1246 Siple time and 1546 Adelaide time. We could not synchronize clocks because international radio schedules were geared to GMT. As Giles put it, 'This thing is bigger than both of us.'

Now we headed north. On reaching the Pacific coast we encountered drifting snow and turbulence. Out to sea along this meridian, there was nothing between us and the coast of Ecuador. Turning inland, Giles headed for a long-abandoned US station named after James Eights, the first American scientist to reach the Antarctic (1830).[1] The station had been in use from 1962 to 1965 and I wondered if anything would still be showing above the snow. We navigated to it by taking a departure from Mount Rex, the nearest nunatak. Fifteen minutes later we came upon two radio masts 6 metres high, all that remained exposed from several buildings and two tall masts. It was a welcome check on our position.

From Eights we headed south to see what might lie in that direction. My chart showed nothing. The terrain dropped away steadily from 360 metres above sea level to about 100 metres, indicating that we had reached an arm of the floating Ronne Ice Shelf.[2] To our right we saw some black specks on the horizon and went to investigate. Sure enough, there were three small nunataks where the map was blank. On the assumption that nobody had seen them before, we were overcome with excitement. Once we had fixed their position, they would be a useful navigational reference point a long way from any other. Moreover it would be wrong, even for glaciologists, to pass

113

up an opportunity to contribute to geological science by bringing home a few specimens.

Circling the nunataks, Giles agreed to land, and a minute later we were sliding to a halt 150 metres from the largest outcrop. Our position was 77°03' south, 78°14' west. Giles left one engine running because, if he should fail to start up again, we would look a bit silly sitting helpless in the wilderness. We had kept in touch with Siple, so they knew where we were. However, the nearest aircraft that could come to our assistance was more than 1,000 kilometres away.

We walked to the rock, climbed it, and photographed ourselves – triumphant. The view from the top was stunning. We were 1,200 metres above sea level and the air was so clear that we could see to the farthest horizon on all sides. Away to the south-west, 240 kilometres from where we stood, rose the giant chain of the Sentinel Range. To the north, east and south there was only a vast expanse of nothingness. It was truly an isolated spot. Geologists require samples of bedrock, not anything loose that could have been carried here by the ice sheet. It was not easy because we had no hammer. However, by exploiting the natural jointing, we broke off enough to give each man all he could carry. Half an hour after landing, we were airborne with 30 kilos of rock specimens.

Back at Siple five hours after starting, we realized that the INS might not have given a precise enough position because three and a half hours had elapsed since it was set. So we refuelled and had a meal before leaving to make a position check on the nunataks. This time I invited Dale Merrick to come along. After reading the INS over the summit where we had stood five hours before, we continued on the same south-easterly heading. Now our quest had become geographical as well as glaciological. Realizing that the ice-shelf boundary was far from where it was shown on the chart, we felt duty-bound to find where it really was. So on reaching the edge of the ice shelf, we followed it.

The boundary was marked by a shallow depression and by tiny cracks caused by the tidal rise and fall of the ice shelf. Because I had sledged over similar terrain before, Giles asked me to do the flying. Ten metres off the ground, banking left and right to follow the meandering boundary, it was a good deal more tiring than cruising straight and level on a compass heading, so after a time I gave up and crossed the end of a peninsula.[3] Coming once again to an ice shelf boundary, I followed it west into a 40-kilometre wide inlet that proved to be 140 kilometres long.[4] This was exciting – we were redrawing the map. At this point I asked Giles to head for the north end of the Sentinel Range to tie down our track on the way back to Siple. When we finally landed, the Twin Otter had flown 2,350 kilometres since taking off just after dinner the day before. Not only had we made significant geographical discoveries but the radio-echo sounder had functioned perfectly most of the way. Nothing was known of ice depths in this whole area before we came.

With our enthusiasm redoubled, we were off the ground again at 1733 on 23 January. This time we headed 154° true with the object of tying down more of the ice shelf boundary, while at the same time exploring an area of potentially very deep ice. After 300 kilometres we climbed over a peninsula on the south side of the long inlet we had seen the day before.[5] Thirteen minutes later we were descending over a steep escarpment, at the foot of which we were again over floating ice. Ahead we could see a low, crevassed ridge stretching across our path. In my notebook I wrote 'Locally grounded area', the result of ice going aground on a seabed shoal. Coming up to it, we followed its convex surface, marvelling as a scene of unremitting chaos flashed by. Never, since landing on Byrd Glacier in 1960, had I seen so many crevasses crammed together so tightly. Even the boldest of mountaineers would have cringed at the sight.

The radio-echo sounder showed that the floating ice shelf was 1,100 metres thick on the upstream side of the obstruction. Forty seconds later as we crossed the grounded area, the ice was only 600 metres thick. The top of the seabed shoal was 500 metres below sea level, but its drag on the ice flowing over it had led to this drastic thinning and the turmoil that we were witnessing. It was a thrilling moment, long to be remembered.[6]

Now we headed for a 160-kilometre-long ice rise that *was* on the map.[7] Faced with a major ice feature like this, I wondered whether Giles would give up low flying over the uphill slope in case the climb proved too steep for the Twin Otter. I watched as the airspeed fell away to 100 knots, but Giles knew that he had enormous reserves of power if the need arose. The rounded summit was 400 metres above sea level; it gave way to a steepening downslope on the eastern side until we came upon a belt of crevasses indicating that we were once again over floating ice. I continued to plot the position of the ice-shelf margin – which we call the grounding line – until we were over the land. On reaching Siple at 2300, we had been in the air for five and a half hours and were well satisfied with the day's work.

Our time was now up because we had used the last drop of fuel that the Americans could spare. They had been more generous than we could have hoped. The next day, well rested, we headed home; but this was no ferry flight. I was determined to squeeze the last ounce of new knowledge from the area. We bade farewell as the Siple crew asked one last favour – that we should buzz them low to provide action photographs of the Twin Otter. We did. Heading east to the big inlet that we had found two days before, I aimed to follow the western margin of Ronne Ice Shelf[8] to check whether it conformed to the existing sketch map. In places it did not. By following close to the land and continuously recording the ice thickness, we obtained enough data to establish the trend of the coastline.

I had been asked by Dr Paul Williams of the US Geological Survey to report on the state of 'Lassiter City', a small field camp on Wetmore Glacier that he had last occupied in 1973. We needed to climb one or other of the Palmer Land glaciers to get to the plateau, so I was happy to oblige. My diary reports:

> A very fine bit of scenery as we ambled up the glacier at 30 ft and 100 knots . . . Shortly after passing the last landmark before the [camp's] location on the map, I told Giles that it would be just around the next corner. We both broke out laughing as we passed right over a 1 ft length of stake, all that remained of the whole camp. Away to the left a few hundred metres up the side near a nunatak was a solitary 3 ft stake with a flag on it. Both stakes appeared to be not more than one inch in diameter. Onward and upward to Mount Vang and then landed Fossil Bluff at 2154.

Spending the night at Fossil Bluff, we had a moment to take stock of what we had been doing. In three days of radio-echo sounding based on Siple we had travelled 5,700 kilometres over areas never before covered. In redefining the earlier speculative boundary of the inland ice sheet, we had added 38,000 square kilometres to the land area of Antarctica while reducing the mapped area of Ronne Ice Shelf by 11 per cent.[9] When we got home, satellite images allowed us to plot the new ice-shelf boundary not only at, but also between, the points where we had crossed it. We had found the thickest floating ice known – 1,860 metres. With the help of my colleagues I later reported these and other findings in the *Philosophical Transactions of the Royal Society.*[10]

The three small nunataks that we had found on 22 January turned out to have an interesting history. They had been seen from the air in 1947 at a distance of about 160 kilometres and named Mount Joseph Haag.[11] However, the reported position was about 70 kilometres from where we had found them. No wonder we thought we had discovered something! The height, I noted, had been overestimated by 1,850 metres. The name was changed to Haag Nunataks in 1977 on the basis of our findings.

The geological results were even more interesting. Bernie Care, a young geologist at Adelaide, identified our samples as 'coarse-grained quartz-feldspar-biotite gneisses' with a dark green basic sill or dyke. Some specimens were later determined by radiometric dating to be a billion years old.[12] This is more than twice as old as the oldest rocks found in the Antarctic Peninsula. The discovery halved the 400-kilometre range of uncertainty in the position of the boundary between two of Antarctica's major geological provinces.

None of this work would have been possible without the collaboration of the US National Science Foundation and Siple station personnel in providing us with accommodation, aircraft fuel and the loan of an INS. Where else in the world, I wondered, would anyone give so generously yet never send an invoice?

During the next month our flights were many and various, providing not a few lessons in the making of a polar pilot. On some days, because of bad weather, it was not safe to fly. Then we would laze on bunks, write up notes or read from the plentiful supply of books.

We resupplied geological and glaciological field parties whenever they asked for assistance. Both these tasks could be combined with ice-depth soundings by adjusting our flight plans. Also, whenever take-off weights allowed, I would offer a ride to someone from one of the sledging parties. I was only too aware that we mortals from the sky were enjoying an unusually privileged existence, wafted over the landscape in armchairs while instruments did most of the work. I knew from my own years of ground travel how they must be feeling. They were invariably thrilled to fly with us.

On one dual-purpose flight we flew east over the Palmer Land plateau to the Weddell Sea. Making a low pass over a lead in the pack ice, we set the pressure altimeter of the aircraft and then sounded several icebergs. After four and a half hours Giles landed on Mosby Glacier beside a sledging party known by its radio call sign of 'Sledge Whisky'. Glaciologist Jim Bishop and his GA Roger Tindley were camped in the middle of the glacier; they had come to take samples for oxygen isotope analysis from different levels in a hand-dug snowpit and also to measure snow temperatures at various depths below the surface. We spent two and a half hours on the ground catching up with their news before heading back to Fossil Bluff.

After that flight, the INS packed up for good owing to an electronic fault, so navigation was back to the slide rule and old-fashioned dead reckoning. For several days we based our flying on Fossil Bluff. On one flight there was a hydraulic problem with the skis and Giles landed wheels-down. We realized what was wrong when we circled back and saw three ruts in the snow. I had often wondered whether, in these circumstances, the machine would cartwheel head-over-heels as the wheels dug into the snow. The aircraft came to a stop faster than usual but no damage was done. We were off again as soon as the problem was solved. Giles must have been having an off day because he omitted to set the flaps for take-off. Hurtling towards the Fossil Bluff rubbish dump, we were still not airborne after twice the normal take-off run. At that point Giles came to and applied flaps. This was one of the very few occasions on which Giles had reason to be embarrassed. However, whenever it happened, I was consumed with admiration for his lightning-fast reactions. In fairness I must add that, given 50 hours or more in the co-pilot's seat, I have yet to fly with a pilot who did not make mistakes. The trick is to correct them before disaster strikes. Giles was good at that.

117

We knew that there was a shortage of ground control for mapping. Fortunately Jonathan Walton, one of the wintering glaciologists at Fossil Bluff, was also a qualified land surveyor. On 31 January we put him down with Tim Stewart beside the most westerly outcrop that we could find. This was at Marion Nunataks on Charcot Island. The place had been seen from seaward by the French explorer Jean-Baptiste Charcot as long ago as 1910, but nobody had ever come ashore. For us, it was a perfect soft landing on a snow col close to the rocks. After unloading the camping party, we took off downhill on a slope leading towards the sea. I had always thought that a downhill take-off would be easy. However, with more weight thrown onto the nose ski, it dug into the snow. By the time the machine came unstuck we were uncomfortably close to the ice cliff. Another lesson learned. Afterwards we laughed about such moments – but never while flying.

Chris Doake and Hugh Macpherson were now operating the radio-echo sounder on alternate flights in order to relieve the tedium of long hours staring into the oscilloscope. Landing back at Fossil Bluff just after midnight, Chris fed us a fine dinner of baked ham with peas and mashed potatoes. It was always a joy to sit round the table discussing the events of the day, generally glowing with satisfaction over the data that we were accumulating.

One day towards the end of a seven-hour flight over Alexander Island, I was flying so that Giles could have a rest. After a while I noticed out of the corner of my eye that he seemed uneasy – Giles very rarely seemed uneasy. Finally he said, 'Charles, I would feel happier if you would fly a bit higher.' Evidently I had been judging our height in relation to the height of sastrugi – we normally did. What I had failed to notice was that, flying into a windless area, the sastrugi had all but disappeared. Cruising at 120 knots, I had unwittingly come down from our usual height to about 3 metres over the snow. We both breathed easier when I came up.

While Giles always did the bulk of the flying, my main task was to free him from worrying about where we were. His habit was to leave me to plan flights of up to seven hours' duration. In the course of the flight he checked from time to time to see that I knew where we were. He would ask for an ETA over the next nunatak, and woe betide the navigator who was more than a minute out. Checking on each other, I suppose, is what the crew up front should always be doing.

Most pilots resent being checked on by a passenger but Giles never did. I asked what I should do if his attention momentarily lapsed and we were heading for the ground. There would be no time to say: 'Excuse me, Captain Kershaw, perhaps you should look ahead.' Giles answered without hesitation: 'Pull up, of course – hard!' There cannot be many pilots who have given their passenger a standing invitation to wrest the controls from their hands. But ever the pragmatist, Giles was right – safety must override injured pride or any other consideration. As it turned out, neither of us ever had occasion to grab the controls.

One day the cabin filled with smoke shortly after take-off. Giles asked me to try to find the source of the problem. I grovelled all over the cabin but failed to locate it. By then the fumes smelled toxic and I reminded Giles that we could land where we were, here and now. This was an unsubtle way of saying that I wanted out. Icy calm as always in an emergency, Giles said that both engines were running normally, so in the absence of flames he would prefer to return to Fossil Bluff. Both of us were glad to get back to mother earth. Fire extinguisher in hand, Bert Conchie helped Giles to track down the problem. They found that the transformer of the 'Gyrosin' compass had shorted out and was burning its insulation. Later I asked Giles how he managed to keep calm in moments when I inclined towards the opposite extreme. He replied: 'Well, Charles, I have had three fires in the air in the last three years. Two were engine fires – much worse!'

We were making considerable inroads into the Fossil Bluff fuel supply, every drum of which had been flown here, so on 3 February we returned to Adelaide and based the rest of our flying there. It was the warmest period of the year and the snow was sticky. On some flights we failed to get airborne on a first run and had to have a second try by following in the ski tracks of the first. When this did not work, Giles sent everyone aft to weigh down the tail so that he could pull the nose ski off the ground.

Although we always started flights in fair weather, things could change in the course of a long flight. Sometimes we found ourselves stuck on top of a cloud layer or trapped beneath a lowering overcast. One paragraph in my diary reads:

> Moderate turbulence gave us an exciting if at times mildly alarming day but the Otter and Giles plod reassuringly on for hour upon hour of buffeting progress. At least, one learns how to handle the machine under the oddest circumstances. At one point we crept between an undulating crevassed glacier and a cloud ceiling 100 feet above it and it occurred to me that some of us do very odd things for a living. The 100 feet was filled only with fleecy whiteout.

Another day we were considerably shaken up by mountain turbulence while passing the Eternity Range. I have always disliked severe turbulence, although the shoulder and lap harness that we wore would hold us securely in our seat even if the aircraft flew upside down. Instinctively, I was hanging onto my seat with both hands. Giles spotted this out of the corner of his eye and quipped: 'Charles, that's cheating!' At moments like this it was reassuring to know that Giles' sense of humour had not deserted him. We had a good laugh and I relaxed – a little.

Turning my head to see how the cabin crew were surviving the turbulence, I noticed that Chris Doake was bent forward with his head in a plastic bag.

He was not prone to suicide, so I concluded that he was preoccupied with disgorging his breakfast. When we landed at Adelaide, he was stretched out on the floor of the cabin. Pedro had been monitoring the instruments.

Leaving them on the ground to recover, we took off with Hugh, this time spending nearly four hours running lines back and forth over the Fuchs Ice Piedmont on Adelaide Island. Brilliant sunshine made it tiring on the eyes. Although we preferred to fly at night, I was aware that the base members, like normal human beings, preferred to sleep then. Everything that we did depended on them. Whenever we were operating, whether in the air or on the ground refuelling, a radio operator and meteorologist had to be continuously on call. Relays of base members took turns at moving fuel drums into place and carrying away the empties. We owed them a lot – we were having the fun while they had the drudgery.

13.   FILLING SPACES (1974–75 continued)

Not lost but 'temporarily misplaced'

There remained some work to do further north. On 20 February Pedro learned from Matienzo station that the weather there was good, so we took off with Chris Doake, Slim Simon and Pedro to do as much work as possible with the fuel that the Argentines had promised. Landing at Matienzo, we dumped our overnight bags and then flew on to Marambio. Marambio – Like Matienzo – belonged to Fuerza Aérea Argentina but it housed a much larger contingent of airmen. Their airstrip was on the flat, mesa-like top of Seymour Island. At the time it was the only bare-ground runway in the whole of Antarctica.

We had arrived at a bad moment. Two four-engined Hercules landed shortly after we did. Each carried a full load of building material, having flown with it non-stop from Buenos Aires. Nobody took any notice of us. When eventually the big aircraft left, Pedro entered into lengthy and sometimes heated discussions in Spanish. My impression was that, as foreigners and civilians, we were less than welcome and regarded with some suspicion. Why had we landed at a military base? Who had told us that we had landing rights? Who had authorized the fuel?

All came right in the end but it took four hours to secure four drums of fuel. We would never have obtained a drop if Pedro had not been patiently cajoling the natives. Eventually we took off, and after one and a half hours of sounding runs, landed at Matienzo for the night. Here the atmosphere was different. Though Pedro still had to do the talking, we were entertained to a good dinner in friendly company. All five of us bedded down on real bunks in a comfortably heated hut.

Next day and in spite of cloud, we managed some good sounding runs over Mount Haddington (1,630 metres). This cone-shaped glacier-covered peak was discovered and roughly charted from the sea in 1842 by the great British explorer Sir James Clark Ross, who named it after the First Lord of the Admiralty.[1] As we approached an Argentine naval station known as Petrel, Pedro advised Giles not to fly low. He was evidently concerned that the navy might be even less congenial than the air force.

In Antarctic Sound we were forced to turn round by angry black blizzard clouds and had to abandon a plan to reach the South Shetland Islands. Landing again at Marambio, we found the base even more in turmoil than the day before. The base was expecting three Hercules visits, while at the same time an emergency appendectomy was in progress at the station's twin-bed hospital. There was much arm-waving but we got away after two and a half hours, having taken on four more drums of fuel.

The weather on the six-hour flight back to Adelaide was dreadful. Giles was forced down by a heavy ceiling until we were hopping over icebergs just off the coast. As we approached the position of Lindenberg Island flying at 30 metres, the visibility ahead dropped to 100 metres. I reminded Giles that the island was mapped as 200 metres high. There was only one thing to do. My diary puts it succinctly: 'So on with the old remedy, open throttles, switch on de-icing, and climb rapidly.' When we reached the top of the cloud layer, it was possible to continue sounding over Larsen Ice Shelf in spite of not being able to see it. The radio-echo sounder added a valuable safety factor to instrument flying, as Chris was able to report whether we were over the sea or the ice shelf. We landed back at Adelaide at 0107, finally creeping into our bunks as quietly as possible so as not to disturb the inhabitants.

Throughout the last few weeks, Bert Conchie had been bringing in the geological and glaciological field parties from far and wide – now the job was complete. The base was bursting at the seams with shaggy and sun-burned faces, all apparently pleased with what they had achieved. We ourselves saw no reason to stop flying before the end of February.

On the twenty-third we did finally achieve my ambition of sounding the South Shetland Islands ice caps. It was three years since David Rowley and I had reached Livingston Island only to be plunged into whiteout. Even now it was a seven-and-a-half-hour flight that would have been impossible without the promise of Argentine fuel. We flew over Palmer station at a respectable altitude, having learned earlier that their resident ornithologist was concerned about the effect of noise on nearby nesting skuas. We had once landed there at the station manager's request to deliver mail. While most of the residents welcomed the visit, an angry ornithologist was over-heard rounding on the station manager for having invited us. Nobody blamed us – at least not in my hearing.

After sounding Anvers Island and Low Island, we continued to Deception Island, noting that someone had laid out a short runway where Carl Ben Eielson and Hubert Wilkins had taken off on the first-ever Antarctic flight in November 1928. On Livingston Island we spotted the camp of a group of BAS geologists led by Mike Thomson. I leant out of the co-pilot's window to drop a note, but as I did so the slipstream snatched an expensive pair of sunglasses from my face. The note was picked up but my glasses were never found.

Following the chain of islands north-eastward, I scribbled a friendly greeting to the Soviet station Bellingshausen on King George Island. Our practice for airdrops was to wrap the note in a strip of red bunting weighed down with a bar of chocolate. In this way notes were invariably welcome. A few minutes after dropping it on the Russians, there was a deafening squeal in our headphones as the Russians tuned into our frequency. Using my very rusty Russian, I had a friendly chat in which the radio operator thanked me for the chocolate and invited us all for a visit. However, there was no time for socializing. Giles asked Pedro to don my intercom set to ask the Argentine stations about their weather.

At this point we headed across Bransfield Strait for Trinity Peninsula. Halfway across, Giles spied a trawler and buzzed her. The crew waved and, as we passed by, I read her name: *Valparaiso*, from San Antonio, Chile. While there was always an element of fun in making a low pass over ships and stations, there was also a serious purpose. Radio conditions were poor and we had difficulty notifying Adelaide where we were. Like the explorers of a bygone age who placed notes in cairns to record their passing, it was our way of making ourselves known. In the event of a forced landing along the way, it would narrow the search area.

In the course of a long and excited conversation with Matienzo, Pedro had learned that Marambio had just closed their airstrip because of fog. Deprived of my headset, however, I was unaware of this. When I recovered the head-set, Giles asked for an ETA. I gave him an ETA for Marambio, where, according to our flight plan, we intended to land. He took it to be an ETA for Matienzo. It was only when the ETA came up that we discovered the mis-understanding. Giles was concerned because the added 200 kilometres to Matienzo would involve eating into his fuel reserve. So we took a short cut over the saddle of Depot Glacier at Hope Bay and thence along Prince Gustav Channel.

Our anxieties evaporated on approaching Matienzo. It was in the clear and our friends were on the airstrip to meet us with four drums of fuel at the ready. My diary records: 'Another very friendly evening with base com-mander Ricardo García and his six genial companions.'

For the last few days of the season we continued operating out of Adelaide and Fossil Bluff, criss-crossing the landscape, running up and down valley glaciers, discovering new features on the ice shelves and filling in blanks where there were no soundings on the map.

Wordie Ice Shelf looked as if it was fighting for survival. Between 1967 and 1972 I had seen big changes in the position of the ice front. Now, three years later, it was obvious that several hundred square kilometres of the ice shelf had simply vanished. Was the ice perhaps returning to equilibrium after an earlier surge advance of the glaciers feeding into it, or was it a response to some subtle change in climate? The full extent of the lost area only became apparent when we compared a Landsat image with the known 1966 position of the ice front.[2]

Wordie Ice Shelf and the mouth of Fleming Glacier in 1966. Retreat of the ice front has since reached the foot of Fleming Glacier. Buffer Ice Rise (centre) has disappeared under the sea

Our final flight was on 28 February to a high snow dome on the backbone of Graham Land near Slessor Peak. We landed at 2,200 metres above sea level to do snow sampling for stable and radioactive isotopes. The air temperature was a warm minus 5 degrees centigrade and there was no wind. Giles had brought along Mike Harris for a day out but now put him to work at the ice drill. Core sampling to a depth of 9 metres took two and a half hours. Meanwhile Giles dug a snow pit for shelter and set to work with a Primus stove borrowed from the aircraft's survival equipment. An accomplished cook, he was soon serving a steaming stew, fruit and coffee.

The snow temperature at the bottom of the drill hole was –16°C. By contrast, the mean annual temperature at Adelaide, which is virtually at sea level, is –5°C.

Taking off after five and a half hours on the ground, I pointed Giles in the direction of a nearby landmark but omitted to state a compass heading.

Choosing the wrong glacier to descend, I sheepishly admitted after a few minutes that we were 'temporarily misplaced'. This is an aviator's euphemism for getting lost. We had come down an eastward-flowing glacier whereas I was intending to fly west. Having guided Giles without a hiccup for some 50,000 kilometres through all kinds of weather and poorly mapped terrain, now on the last day of the season I had let him down. My embarrassment lasted some time before I once again gave ETAs that made sense. We landed back at Adelaide nine hours after leaving. Apart from my blunder, it was a magnificent last flight.

The next day, 1 March, my diary gloated: 'Overcast and snowing, but it does not matter!' There was very much an end-of-term feeling at the base. Chris and Hugh took our equipment out of the aircraft and Pedro developed the radio-echo films. After two days, both aircraft were packed and fuelled ready for their long ferry flight back to Canada.

For us it was a moment to take stock. We had been in the air for an average of 6.7 hours on each of 32 days. In the space of 6 weeks Giles had kept us in the air for 214 hours, more than in all our earlier radio-echo sounding seasons combined. The total distance flown was greater than the circumference of the earth. We had studied vast tracts of land where nothing was known about the thickness of the ice sheet, and we had identified areas of critical interest for future studies on the ground.

At the same time we had landed ourselves with a colossal task of data reduction to make the results accessible to others. The season's work, I felt, had achieved the kind of exploratory framework that I had set out to build eight years earlier. Future radio-echo sounding should concentrate on particular areas where we might best contribute to an understanding of the mechanics and dynamics of the ice sheet.

Much of the knowledge we gained had a bearing on research projects that BAS glaciologists have pursued over the years since 1975. In some areas, such as Alexander Island, our network of soundings had become so dense that it led to a new kind of map showing ice thickness isopleths (contours) at 25-metre intervals.[3] Our interest in Ronne Ice Shelf led to continuing projects with participants from BAS, Germany, Norway and Russia.[4] The finding of the thickest floating ice in the world led to years of productive work on Rutford Ice Stream.[5] This in turn sparked interest in the new field of sub-ice shelf oceanography.[6] My own work in the glaciological interpretation of satellite imagery led to a volume containing examples from many parts of Antarctica.[7] Finally, our brief sighting of a smooth patch of bare ice at Patriot Hills led to the development of ice airstrips for transport aircraft.[8]

With some questions of geography settled, the place names could be clarified. Looking at maps of the Ellsworth Mountains and the Antarctic Peninsula some years later, I was startled to see so many names of my contemporaries. Excluding those which I had proposed, I found more than 150 places named after individuals whom I knew. If I am accused of name-dropping, I shall plead guilty. I do it here to underline just how much of the exploration of Antarctica has taken place within a lifetime. In 1975 we were still at it.

John Biscoe arrived offshore on 4 March and in the space of three hours we had everything on board for the voyage home. On the way north we were able to make a brief goodwill visit to the Chilean station Presidente Eduardo Frei that sits cheek by jowl with the Soviet station Bellingshausen on King George Island. Both stations were put there because there is a good anchorage and a large area suitable for building. However, it has always seemed strange that on a continent much larger than Europe, two nations should be competing for space on one short stretch of beach. There was a potentially delicate situation in the event of a clash of interests because the two sponsoring governments had severed diplomatic relations at the time of the overthrow of Chile's Marxist President Salvador Allende in 1973. Chile's new President Augusto Pinochet had vowed to 'exterminate Marxism'.

Against this background the two station commanders had made peace and now consulted each other on all areas of potential conflict. Together with *John Biscoe*'s Master, Captain Malcolm Phelps, and a delegation of Fids, I called on the Russian base commander, Nikolay Ovchinnikov. We came bearing gifts, as is customary in the Antarctic: a case of Scotch whisky, a selection of British Antarctic stamps and some small pennants. These were well received and we were plied with vodka, biscuits and pickled fungus – a Russian delicacy. One of two telephones on the commander's desk was red. Enquiring what it was for, he replied, 'That's the hot line!' It was for direct communication with the Chilean base commander. By this means they kept the peace – and still do today.

In Stanley four days later, Malcolm Phelps and I were entertained to dinner by the Governor, and the following day for a supper-cum-film party with staff members from the ship and some local dignitaries. Two days later we reciprocated by entertaining 50 guests at a large cocktail party on board. Among the guests were two attractive young sisters from Argentina. They had come for a year to teach Spanish to Falklands children. It was part of a policy supported by both governments to bring the two countries closer together.[9]

Another recent development with obvious political implications was an air service connecting Stanley with the coastal town of Comodoro Rivadavia in Argentina. This was not operated by any civil airline but instead by the Argentine military. Linea Aérea del Estado (LADE) ran a twice-weekly service to the mainland with a 30-passenger Fokker F27 aircraft. The crew were serving members of Fuerza Aérea Argentina. They took advantage of the fact that liquor in the Falkland Islands was cheaper than at home. My small party secured places on the 17 March flight. Every seat was filled, and

Jonathan Walton

Twin Otter approaching Fossil Bluff

It took intense concentration to fly this low

A tight squeeze between icebergs

Halley Bay. The station is deeply buried by snow that has fallen since the huts were built on the surface years before. The author (in blue) with Bill Sloman

A visit to Palmer station

We dedicated a memorial cross to three climbers who died on Mount Peary

Midnight visit to the American camp in the Ellsworth Mountains. Left to right: the author, John Splettstoesser, Bjørn Andersen, Bob Rutford and George Denton

the crew virtually barricaded themselves into the cockpit with a mountain of cases of whisky. The aircraft was so heavily loaded that we only left the ground in the last 50 metres of the runway.

In Buenos Aires we were met by René Dalinger, the dean of Argentine glaciologists, and by Pedro's wife Mariana and her three sisters. The next day I visited Instituto Antártico Argentino to meet the other two kingpins of Argentine glaciology: Arturo Corte, whom I had come to know in Sweden 26 years earlier, and Benito Colqui. My interest was to encourage, as far as possible, Argentine scientists to play a fuller part in what was now becoming a major field of research. However, as the institute was – anomalously – part of the Department of Defence, they held out little hope that the management would be interested. Dalinger had recently vacated the post of Chief Scientist in order to devote his time to furthering the cause of Argentine glaciology.

A better bet for research support was Consejo Nacional de Investigaciones Científicas y Técnicas (CONICET). Its Director, Vicente H. Cicardo, invited me to visit parts of the Andes to see, and perhaps advise, on the use of glacier run-off for irrigation and domestic water supply. I was totally unqualified in this field and also undeserving – but I accepted the invitation. Colqui, with a twinkle in his eye, said that much of the run-off was used in vineyards, so it would be essential to follow the whole hydrological cycle all the way from its source in glacier ice to the ultimate destination in a bottle. It was perhaps not a coincidence that Dalinger and Colqui were keen to come with me as guides.

The next day I was taken to Servicio Meteorológico Nacional to discuss what could be done to improve weather forecasting in the Antarctic Peninsula. The Director was another military man: Comodoro José Eugenio Echeveste. Colqui was head of the snow and ice section. A bronzed and handsome 56-year-old, Colqui came from aboriginal Andean stock. Like Dalinger, he was a geologist by training but had obtained his Ph.D. through studying glaciers flowing into Lago Argentino. Afterwards he spent three summer seasons doing glaciological work in the Antarctic. We had long discussions about what, collectively, we had achieved and how much more remained to be done.

I had with me some American weather satellite images of Antarctica and Echeveste eagerly asked how I came by them. Although his need was surely greater than mine, I was surprised to find that he did not know how to receive them regularly.

I also showed some NASA Landsat images of parts of Antarctica and they too aroused great interest. I was summoned to appear before the Dirección Nacional del Antártico. Ushered into their boardroom, I found myself facing 15 senior officials, half of them in uniform. Without so much as a minute's notice, I was asked to give an impromptu lecture on the use of satellite images in Antarctic mapping. Although some of my North American colleagues had led the way in this kind of mapping, BAS had followed close behind.[10] It was obviously the way to go but we were only at the very beginning. My hope was that the Argentine Government would share in the task of mapping parts of the area.

The audience seemed transfixed and kept me going for an hour. I faced a barrage of questions and finally put the cat among the pigeons by showing them a satellite image of the part of Filchner Ice Shelf on which there was an Argentine army station – Belgrano. We could see a rift in the ice shelf advancing directly toward Belgrano from the east. Sooner or later it could set the station adrift on an iceberg in the Weddell Sea. They asked how soon this could happen but I declined to guess.

After another round of visits, Dalinger and I flew to Mendoza at the foot of the Andes. We were met by Juan Carlos Leiva of Instituto Argentino de Nivología y Glaciología (IANYGLA). Dalinger had set up the institute some years earlier. Arturo Corte was now the Director but he had remained in Buenos Aires to attend meetings. After a night in the best hotel, we were shown over the institute. As in Buenos Aires, I was asked for an impromptu lecture. However, I was delighted to find that here they were already using satellite images to make a glacier inventory of the Andes. It was a monumental task that could take a decade but they had made a start.

Mendoza is a clean and spacious colonial city with a gorgeous backdrop of the Sierra de los Paramillos. It was colonized from Chile in 1561 but it was from here that, in 1817, General José de San Martín set out to cross the Andes to help in the liberation of Chile. In 1861 the city was destroyed by an earthquake and fire with heavy loss of life. The modern city has a population of 600,000, with mostly low buildings as a precaution against further earthquake damage.[11]

Dalinger and I set out to climb the Andes – not on foot but by taking a taxi. East of Mendoza there is semi-desert, but west toward the Andes there are fruit orchards and vineyards. The road continues to Santiago, Chile, but we were heading only for the watershed, which is the international border. It was a spectacular winding road past tiny settlements with names like Uspallata, Punta de Vacas and Las Cuevas. At Puente del Inca there were hot springs and the remains of a hotel that had collapsed in an earthquake.

We stopped for a picnic lunch at a point where there was a good view up a stream valley to the towering summit of Aconcagua, the highest peak in the western hemisphere (6,960 metres). A small graveyard faced the mountain and held the remains – or at least the headstones – of climbers who had lost their lives climbing. Our driver drew a pistol from his shoulder holster and spent some minutes firing at rocks. Nobody remarked on this, so I assumed that keeping his gun in order was as important to him as keeping his taxi meter working.

The international border at 3,854 metres above sea level is marked by a gigantic statue of Christ the Redeemer. A modern tunnel cuts beneath the watershed, so the highest point on the road is at 3,300 metres. We took a short walk to a point where we could see a distant glacier. This was the sum total of our day's 'work'. Serious studies of the ultimate destination of the glacier run-off were to begin later.

The conversation turned to politics. The Chilean border guards, I was told, have been known to shoot hitch-hikers crossing the border from Argentina without papers. I refrained from adding that in Chile I had been told of eastbound hitch-hikers being dealt with in the same way by Argentine border guards.

Back in Mendoza, we spent the evening discussing the merits of some local wines. In the morning, Dr Leiva drove us through patches of desert and oases to San Juan, a small city 170 kilometres north of Mendoza. San Juan was destroyed by an earthquake in 1944 with heavy loss of life. The rebuilt city is clean, tidy, and well-planned. This is the centre of the wine country. We were met by Benito Colqui, who ushered us into a plush office to meet Ingeniero Mario Alberto Gallo, Director of Departamento de Hidráulica. On one wall was a large oil painting of Juan Domingo Perón in an ornate frame, facing, on the opposite wall, a marginally smaller portrait of Eva Duarte Perón. On an adjoining wall there was a crucifix. The Director opened a second door and in came several pressmen and a television crew. Not again, I thought. Luckily Colqui did most of the talking.

Once more I was asked to lecture off-the-cuff, with Colqui translating. I felt like a fraud. Yet such were the feelings of isolation of these people that they chose to lionize a visiting foreigner who had almost nothing to contribute to their lives. It was embarrassing – but they hung on every word.

After a night's sleep we drove to a large bodega to sample several varieties of wine. It was excellent and I asked why there was not much of an export trade to Europe. 'We have more sense,' the Manager said. 'We keep the best wines for Argentina and export what is left over.' In an aside, he said that unfermented grape juice was exported to France, where it was bottled and sold to undiscerning consumers as 'Produce of France'. I sensed that the British were considered to be in that category.

Colqui told me that the third largest immigrant group in Argentina after the Spanish and Italians were the Lebanese. We were entertained to lunch at the Lebanese Club by Dr Indalecio Carmona Ríos, President of Sociedad Científica Argentina – Seccion San Juan. Another guest was Dr Manuel Alejo Vera Correa, vineyard and winery owner. Colqui leaned across the table and confided to me that his mother had been a domestic servant on Correa's estate.

Colqui was very proud of his aboriginal origins. Without shoes in his early years, he had begun as a labourer wielding a pickaxe in a quarry. With the proceeds from this work he went to university and eventually became a teacher. He said that, as far as his Quechuan origins were concerned, he had never encountered racial prejudice in Argentina. Coming from mountain country, glaciology had been his hobby. Now, apart from working in glaciology, he was Rector of St Alban's College, a private school with 200 pupils, only 4 of whom were girls. Thirty per cent were of British descent or parentage. Half of the instruction was in Spanish and half in English.

Another winery we were shown was in full swing, with truckloads of grapes coming in at one end and juice being pumped into railcars at the other. Some of the storage tanks were so massive that the place could be mistaken for an oil refinery. Giant casks in the cellar were made of French oak. The vines too originally came from France. I was presented with a bottle of 15-year-old brandy to carry home.

The following day we were given a chauffeured official car from Departamento de Hidráulica and driven to flumes and water-filtration plants in the foothills. There was a lot more water here than at Mendoza. We spent the night in a government rest house and set off early in the morning to drive up the Agua Negra valley. Leaving Las Flores, the road climbed via a series of zigzag bends to a Customs post and finally to a pass at an altitude of 4,765 metres. This was Portillo del Agua Negra on the Chilean border. Alighting for a short walk, we rapidly began to feel the altitude, puffing and blowing like sick men. It was desperately cold and windy.

Some peaks in the Cordillera de Olivares on the south side of the road rise to 6,215 metres. On top of the pass we stopped to look at a memorial to the dozens of car and bus drivers – not to mention their passengers – who had skidded over precipices. In places there were patches of wet snow on hairpin bends, and to my way of thinking our driver seemed bent on adding another plaque to the memorial.

Lower down the valley we followed a concrete aquaduct that supplied thousands of hectares of vineyard. After a relaxing half-hour in a hot mineral bath by the roadside, followed by a siesta in the rest house, we were returned to San Juan. It was Good Friday, 28 March, and we came just in time to enjoy the Easter pageant.

On most of these excursions the car was stopped at intervals for *Yerba mate* (Paraguay tea). Communal mate drinking is an ancient ceremony reminiscent of passing a pipe of peace. While we sat cross-legged on the ground in a circle, boiling water was poured into a small, silver-mounted calabash holding the tea in the form of a coarse powder. A little sugar was added and the infusion then passed round, each of the group drawing it through the same *bombilla*, a reed. It does not do to think about hepatitis because great offence would be taken if anyone drank through his own straw.

My short stay in San Juan brought to an end a most memorable Argentine odyssey. I had made many friends, but knowing that this corner of South America was far from my usual itinerary, it was a sad parting. Indeed I never saw René Dalinger or Benito Colqui again. Colqui died in 1977, Dalinger in 1979.

14. GRAND TOURS (1976–77 and 1979–80)

A lesson in diversity

Shortly before the 1974–75 Antarctic field season I had been given a new job as Head of the BAS Earth Sciences Division. This meant overseeing work in geology and geophysics as well as glaciology. One day I was responsible for a scientific staff of 11; the next day it was 39. Most of the staff were at the start of their careers, working on five-year contracts, and evidently in need of supervision. I could not safely handle such a broad spectrum of science, so my first task was to delegate responsibility through section leaders to oversee particular aspects. This involved progressively increasing the ratio of permanent to contract staff. Fortunately that was made easy by having some very competent staff to promote.

Challenging and exciting though the job was, the workload was such that I could no longer launch myself into fieldwork whenever the previous season's results were on the way to publication. I concentrated on launching others.

Sir Vivian Fuchs had retired in 1973 after serving as Director of BAS for 26 years. The new Director was Dr R.M. (Dick) Laws, a zoologist whose early fieldwork had been with BAS. Luckily for all the management team, he understood that directing men in the field entirely by remote control would not be conducive to mutual respect between men 'at the coalface' and their supervisors half a world away.

My first opportunity to go south after the 1974–75 season came in 1976–77. Feeling slightly handicapped in my new post by my former specialization, I asked to visit all the BAS research stations in order to see how my own division's work compared with the other two: Life Sciences and Atmospheric Sciences. The three divisions were roughly equal in terms of manpower.

I joined *Bransfield* in Montevideo on 10 December 1976. John Cole was Master, but now the job was shared, John doing the first part of the season and Stuart Lawrence the second. Both men had served for many years in BAS ships. However, it was no longer reasonable to expect them to work at sea without home leave for eight months of each year.

After the usual round of entertaining and being entertained at Stanley, we visited all the BAS research stations in turn. The first was Grytviken, a biological research station on the island of South Georgia. I had been here in 1949 when the place was a busy whaling station. Next came Halley Bay, on the Weddell Sea coast, a geophysical and atmospheric sciences station;

Signy, in the South Orkney Islands, another biological station; Bird Island, a third biological station; Argentine Islands (later renamed Faraday); and Adelaide. On the way we called at Punta Arenas, Chile; Almirante Brown, an Argentine station in Paradise Harbour; and the US Palmer station on Anvers Island. After completing this tour I had a much better understanding of BAS's very broad range of activities.

Flying was still based at Adelaide, but was to be moved at the end of the season to a new station under construction at Rothera Point – the site that Alan Smith and I had reconnoitred in January 1975. The Adelaide runway was becoming dangerously icy and, although it was still usable, the new site offered snow runways of unlimited length and also the possibility of constructing a hard (bare-ground) runway in the future.[1]

Unfortunately, the season was overshadowed by two sad events. Three lives were lost on a climbing expedition on Mount Peary near BAS's Argentine Islands station in September 1976. The climb had no connection with any scientific programme. A policy of long standing had been that scientists living for two years at a stretch in Antarctica deserved to be allowed some time for recreation. The three climbers had prior experience and were well equipped for the conditions. They achieved their goal and radioed from the summit. When they failed to return, a ground search was launched and the Argentine Marambio station, whose aircraft remained in Antarctica throughout the winter, was asked to help. One aircraft flew round the mountain and saw two pairs of skis below the summit but no sign of life. Bad weather prevented further searches until the elapsed time had exceeded human endurance.

On 9 February 1977 we held a short dedication service at a memorial cross erected on nearby Rasmussen Island. It was snowing hard as a party from the ship, and all those who had worked with the lost climbers, doffed their hats and meditated on the price that we sometimes pay for pitting ourselves against this unforgiving land.

The other sad event was the crash of one of the two Twin Otters in a whiteout on 21 January 1977. It was VP-FAP, the aircraft that had served Giles Kershaw and my small group so well two years before. Luckily there was no loss of life, though the aircraft was a write-off. The second BAS aircraft had been leased to the US Antarctic Research Program and was at the time operating on the Ross Ice Shelf with Giles Kershaw as pilot. After a marathon rescue flight across Antarctica by way of Siple station, a distance of 3,000 kilometres, Giles landed at the crash site 21 hours after the accident. Using their on-board survival equipment, the five men from the downed aircraft had followed all the proper emergency procedures and were found camped beside their machine. Good training pays off.

***

During the following three years I managed to build up the staff of the Earth Sciences Division to 47. The number does not take account of the numerous staff in other departments without whose efforts we would never get anyone to the Antarctic. We were totally dependent on the smooth operation of RRS *Bransfield*, RRS *John Biscoe* and the Air Unit. Of the 11 countries maintaining wintering stations in the Antarctic in 1976, Britain was unique in having all of its staff – in science, logistics and administration – under the same management. Largely for this reason, BAS was and still is the most cost-effective research organization involved with Antarctica.

My next grand tour was in the 1979–80 summer season. Our field operations were to be inspected. BAS had resisted a number of attempts to inspect us at close quarters because of the danger of creating 'instant experts' who might expound on Antarctic policy after seeing only a small part of the enterprise. We knew that it took a long acquaintance with BAS to understand the complexity of working in a hostile environment at the other end of the world. In the government administrative hierarchy, BAS now came under the Natural Environment Research Council (NERC), based in Swindon. NERC derived its funds from the Ministry of Education and Science. Our inspectors were to be Dr Peter Twinn, Second Secretary of NERC, and Derek Gipps, Head of NERC Building Services. Derek was formerly on the BAS staff, so in his case we were less afraid of snap judgments. I was asked to accompany them southbound as minder.

Jim Parker, the Governor of the Falkland Islands, had been primed to play his part. On flying to Stanley from Argentina on 7 December 1979, we were met by Miriam Booth, BAS liaison officer, and then whisked away to Government House by the Governor's chauffeur. We were ushered into bedrooms each furnished with everything we needed, including a full bottle of gin and another of whisky. I idly wondered if this was a character test – perhaps the fluid level was recorded daily for entry on our files.

Jim and his delightful wife Deirdre made ideal hosts. We had much to discuss about developments in the Falklands and Antarctica, and kept at it nonstop from teatime until well after midnight.

The next morning His Excellency looked the part. It was the annual parade to celebrate the Battle of the Falkland Islands on 8 December 1914, in which a British squadron routed a squadron of German cruisers approaching the islands. The Governor was dressed in all his finery – tail-coat, sword and lots of gold braid. To accommodate the ostrich feathers on his hat, the official car was built as a London taxicab. Royal Marines of the Falkland Islands Defence Force were on parade in their best uniforms.

A day later we were being tossed on the high seas in *Bransfield*. Stuart Lawrence was Master for the first half of the season. After an informal visit to Esperanza, an Argentine army station on Trinity Peninsula, we moved to Palmer station to put ashore eight American passengers whom we had carried south at the request of NSF. We were shown over the station by their radio technician, John Wells. Peter Twinn was impressed by the opulence of the station. I was mindful that in due course he would note that, by comparison, Rothera was a spartan outpost of empire.

In order to speed Twinn's and Gipps' inspection, I had arranged for Garry Studd, BAS's chief pilot, to fly to Palmer in a Twin Otter to pick us up. We put Twinn in the co-pilot's seat to enjoy the landscape and to learn from Garry about the aircraft and their uses. On arriving at Rothera two hours later, we were whisked into the second aircraft to fly to Fossil Bluff. This was a snap decision to take advantage of good weather. Because bad weather is all too common, even some BAS headquarters staff had never seen Fossil Bluff on their brief visits to the Antarctic. Twinn was lucky.

Fossil Bluff, at the end of the logistic lifeline and supplied entirely by air, was a good place to start. Every kilogram flown there has to be justified, and there are no frills. Four men have sometimes wintered together in the one-room hut.

After returning to Rothera, the visitors spent a full day inspecting the place before flying back to Palmer to rejoin *Bransfield*. I stayed at Rothera, my duties as inspectors' minder at an end. Shortly afterwards, I flew back to Fossil Bluff with Garry. At 'the Bluff', as Fids call it, we were pinned down by poor weather until Christmas. Garry and I took turns at cooking, washing up and fetching water from a nearby melt lake. Our Christmas celebrations were uncomplicated and delightfully relaxed. Most of us had presents from home – a poignant reminder of our isolation in this place. Dinner consisted of roast chicken, roast potatoes and runner beans. Garry baked a fine cake, fully iced and loaded with rum. The beer ration was one mug each – nobody needed more. The next day I baked bread and made lasagna while both aircraft flew fuel loads to points south.

A party of glaciologists had planned a series of experiments on Rutford Ice Stream, one of the glaciers far south of here that we had flown along when radio-echo sounding five years earlier. It was 930 kilometres from Fossil Bluff, making it a round trip of 1,860 kilometres, which was beyond our range. Anticipating this difficulty, Garry had established a fuel depot at the Sweeney Mountains – roughly halfway. By refuelling at this intermediate stop, the Otter could reach well beyond its normal range. We called this leapfrogging.

It looked clear to the south for the first time in a week. Simon Stephenson (glaciologist) and Ric Airey (GA) piled all their camping gear and survey equipment into the aircraft. After they themselves were on board, the only place left for me was on top of the load. Lying prone, I could scarcely see out, but we were not there to enjoy the scenery. I was only a passenger, or 'tourist' as we used to call summer visitors, but with a vested interest in the success of the operation. We took off at 0220 on 28 December 1979, heading straight into the sun.

After the refuelling stop, Garry flew south over solid stratus for two hours until suddenly the whole sweep of the Ellsworth Mountains appeared, clear from top to bottom. It was a scene to cause a sharp intake of breath.

We planned to visit a large American camp at Welcome Nunatak to discuss future collaboration. As usual, we had forgotten that normal people sleep at night, so our welcome was muted. Indeed it took some minutes for anyone to emerge from a row of Jamesway huts. When they did come out, I recognized four old friends, all of them professors of geology: John Splettstoesser, Bjørn Andersen, Bob Rutford and George Denton. Luckily, the joy of meeting overcame any resentment they felt at being woken from a sound sleep. They recognized that it was our only opportunity to visit.

Three large helicopters, all brought from McMurdo in LC-130 aircraft, were neatly parked in line abreast. We knew that our own geologists would be consumed with envy at the freedom of the Americans to say exactly where they wanted to go – and half an hour later to find themselves standing on the very spot.

I had been planning a combined BAS/US geological programme for some future season, so I explained my proposals while they explained theirs. Evidently our plans would dovetail admirably. Since night was not the time for socializing, we left the locals to their interrupted slumber.

Thirty minutes later Garry landed on Rutford Ice Stream – named for one of the men I had just been talking to. In contrast to the Welcome Nunatak camp, this was to be a British-style camp – two pyramid tents, two skidoos and no helicopters. But as our men had no need to climb mountains, helicopters would have been superfluous. They were studying the ice and needed to travel over it.

Leaving Stephenson and Airey with the load we had brought, Garry and I climbed through cloud to 3,500 metres and had an uneventful four-hour non-stop flight to Fossil Bluff. There we slept for two hours, refuelled the aircraft and returned to Rothera. I was in need of a longer sleep but Garry – always the enthusiast – took of on another mission.

The next day I was suffering from back strain after lifting heavy boxes – 'Fid's back' in the local idiom. Falling short of his best bedside manner, the station doctor told me, 'You have osteoarthritis – everyone of *your* age does.' Having worked in the Antarctic since before he was born, I swallowed the brickbat without protest.

An unusual event was the arrival on New Year's Day 1980 of a Twin Otter belonging to the Chubb Fire Company and piloted by Giles Kershaw. Gerry Nicholson, his engineer, was also an ex-Fid. Giles had retired from BAS the year before and was now flying in support of Sir Ranulph Fiennes' TransGlobe Expedition,[2] an entirely private affair. Indeed until the expedition became inevitable, it had been opposed by governments on the grounds that it would expose them to the real possibility of search-and-rescue missions.

I knew from my own experience with Giles that, if he believed that something was possible, there was no stopping him. He had flown direct from Stanley – the first non-stop flight between the Falkland Islands and Rothera. After five days' delay owing to bad weather, Giles and Gerry continued to Halley, a six-and-a-half-hour flight over the Weddell Sea. The same day we were quietly celebrating the results of our new weather satellite picture receiver, a home-made affair built by Dick Kressman. Some hours later, Garry was in the radio room studying the first pictures. Giles had reported his position over the radio. Garry, without revealing how he knew, told Giles and Gerry, now 1,500 kilometres from Rothera, that they were just coming out of cloud, that they had open water below and would see a cluster of icebergs to their left. Giles was baffled – but not for long.

As part of a plan to monitor the stability of the ice sheet, I arranged to fly to Temnikow Nunataks on the Palmer Land plateau to repeat a levelling survey that Jim Bishop had done five years before. He had made precise measurements of the level of the snow surface between two nunataks 670 metres apart. Now we could discover whether there had been any change and if so how much. Geoff Somers agreed to accompany me as GA and assistant surveyor.[3] On the way, we landed on George VI Sound to unload a heavy ice drill that was to be used for penetrating through the ice shelf in order to measure the temperature and salinity of the sea beneath.

On leaving the place, I was sitting quietly in the co-pilot's seat waiting for take-off when Garry said: 'Take-off checks complete – you do it!' I thought back to my first solo flight in 1960 when, without warning, the instructor got out of the aircraft and said, 'All right, off you go!' Now Garry was giving me no time to think about it. The aircraft was lightly loaded. Opening the throttles for take-off while watching the torque meters, I was disconcerted to look up seconds later to find that we were already airborne. Now I knew that, following osteoarthritis, the next diagnosis would be cerebal decay. Garry was too polite to comment.

Garry put us down 100 metres from the largest of several small nunataks. My diary reports:

> Dingle clear. Beautiful views down big glacier to east coast. Eland Mts spread out before us . . . Set up tent in peace. Geoff pitched it better than I have ever seen a pyramid pitched. Lovely supper of meat bar stew, biscuits, butter, cheese, tea, cocoa.

In proper English, dingle means a small wooded valley, but in Fids-speak it means good visibility and a cloudless sky.

The following day the temperature was –19°C with a 20-knot wind. We skied around looking for the bench marks that Jim Bishop had etched into bedrock, and luckily found them without difficulty, but it was too windy for survey work.

136

Because of high winds it took us three days to run the survey line out from one nunatak and back to the same point. The closing error was 6 millimetres. Evidently some skills had not deserted me. The snow surface level proved not to have changed at all. We concluded that although average air temperatures on the coast had risen slowly over the years, the effect had not yet reached this far inland.

We were now ready to be picked up. The daily radio contact between all field parties and Rothera is known as the Goon Show. Everyone listened to everyone else's sitrep (situation report). One of the two aircraft had broken down and Garry was flying to Punta Arenas to pick up spare parts. We must be patient. We read books and then exchanged them. I had brought Roland Huntford's *Scott and Amundsen*, Michael Holroyd's *Lytton Strachey*, Parkinson's *Mrs Parkinson's Law* and *The Law and the Profits*, and Armstrong, Rogers and Rowley's *The Circumpolar North*. We went skiing and Geoff built four igloos, each one an improvement on the last. We slept in one to see what it felt like. Some days the weather was so calm that we could sunbathe topless but on other days we had winds of 30 knots.

There had been opposition in BAS headquarters to the deployment of field parties that had no means of escape other than by air. We were in just such a position. We knew that aircraft can break down or worse. However, having gone hungry in my younger days, this time I had brought full rations for 75 days. In addition, Geoff had found seven cases of dog food in an old sledging depot 3 kilometres away. We concluded that we would survive until about midwinter, giving plenty of time for a rescue party to sledge in from Fossil Bluff.

Garry came for us on 24 January 1980, ten days after we had finished the survey. On one side of the camp there was drifting snow 3 kilometres away, and on the other side banks of fog, so we might have spent longer if he had not come at the right moment.

I spent the next few days visiting geological and glaciological field parties in the course of resupply flights. One party that we moved between camps jammed two skidoos, two 4-metre-long sledges, 20 boxes of food and rocks, with so much other equipment that the men themselves had to lie prone on top of the load. American airmen call this being 'cubed out', in contrast to 'grossed out' to the aircraft's maximum weight.

Back at Rothera I was asked to give a lecture on 'BAS in the Heroic Era', which meant my first season with BAS in 1966. My colleagues evidently had some difficulty relating to anything before that.

The US Coast Guard icebreaker *Polar Sea* got in touch on 26 January to say that they were to arrive the following day with an official party to make a formal inspection of Rothera under the terms of Article VII of the Antarctic Treaty. The purpose of these inspections is to ensure that the provisions of the treaty are being observed. Any party to the treaty can inspect

Geoff Somers and the igloos he built to while away the time

Geoff Somers (hidden by steam) and the author in their igloo

the stations, ships and aircraft of any other party 'at all times'. Up to 1980 about 100 stations had been inspected – more by the US than any other state. The Russians had never sought to make inspections.

Courtesy generally requires giving 24 hours' notice and this was all we were given. My diary reports, 'Great excitement at the idea of helo [helicopter] visitors. . . Everyone tidying up.' Two helicopters arrived with a seven-strong team led by Tucker Scully of the State Department. Al Fowler, an old friend from NSF, came along for the ride. One of the inspectors was Darold Silkwood of the Arms Control and Disarmament Agency. Article I of the Antarctic Treaty prohibits the 'testing' of any type of weapons. Rothera had two rifles which were used for killing seals for dog food. Although often fired, they were never *tested*, so we were within the law.

I had no need to wait for *Bransfield*'s arrival two months later, so I asked *Polar Sea* whether I might hitch-hike with them to Palmer station. To this they agreed. However, I had another favour to ask. The geologists wanted a food and fuel depot to be put on James Ross Island at the north end of the peninsula to provide for work in future seasons. *John Biscoe* was on her way to Palmer to pick it up. If *Polar Sea* could take the depot from Rothera to Palmer, it would save us a Twin Otter flight there and back. I broached the subject with our guests, who asked how heavy it was. They showed amazing forebearance on learning that their hitch-hiker's baggage weighed 600 kilogrammes – 30 times as much as the normal allowance on an airliner. The ship was 20 kilometres away, and that meant two extra helicopter flights.

We invited the inspection party in for a roast beef dinner. Now refuelled, they were led by Mike Sharp, the Base Commander, on a very thorough tour of the station. Whereas 20 years ago the inspectors might have been more interested in searching for hidden armaments, now the priorities had changed and they were interested in environmental matters. How did we dispose of sewage, food, batteries and rubbish? They were given the answers and went to check for themselves.

On declaring themselves satisfied, we led them to the bar. It was now late evening and they were quite ready to inspect – and sample – the bar supplies. Considering what they were about to do for us, that was the least we could offer. The helos finally came at 2230 to take them – and their hitch-hiker – back to the ship. Donning one-piece watertight immersion suits, we were whisked away, in total darkness, to a floodlit ship already under way.

On landing and scrambling out of the immersion suit, I was led to the bridge and presented to the CO, Captain Herbert H. Kothe. He explained that there were about 160 people on board and therefore no available bunks. Would I mind sleeping on the couch in his day cabin? I felt lucky to have anywhere to lay my head. It was now well after midnight, so I was asleep in no time.

Next morning the Executive Officer, Commander B.D. Lovern, took me

139

on a guided tour of the ship. There seemed to be people everywhere but every compartment was spotlessly clean. In addition to the crew and inspection team, there were five scientists on board with access to well-equipped oceanographic and biological laboratories.

Polar Sea and her sister ship *Polar Star* are the world's most powerful non-nuclear icebreakers. Only the Russians operate nuclear-powered icebreakers. With a length of 122 metres and displacement of 13,000 tonnes, *Polar Sea* is a vast and complex piece of machinery. Three 5-metre diameter controllable-pitch propellers are powered by a diesel-electric system developing 18,000 shaft horsepower. This gives a ratio of shaft horsepower to displacement tonnage of about 1.4, which is not particularly high for a modern icebreaker. However, *Polar Sea* has a formidable weapon up her sleeve. In heavy ice, gas turbines totalling 60,000 shaft horsepower can be put to work. Similar to aircraft jet engines, the turbines burn fuel much faster than the diesels and so are only used when really necessary.

I was invited to dine with the Captain in his stateroom. Other guests were Lieutenant Commander Maria Kazanowska from the staff of the Chief of Naval Operations, Commander Lovern and a helicopter pilot. A daily sitrep showed that the total of US personnel on land in the Antarctic this day was 922 – with 768 of them at McMurdo station. Adding ship's crews would perhaps double that number. No other country kept comparable numbers in the Antarctic.

The following morning we steamed into Palmer while the inspection team flew off to visit the Argentine station Almirante Brown in Paradise Harbour, 60 kilometres to the east. Together with my 600 kilos of baggage, I was taken ashore in a landing craft and introduced to the Station Manager, Karl Wendelowski. I asked if he would consider feeding and housing me until *John Biscoe* arrived. Karl had been carried from the US to Palmer by *Bransfield*, so he was favourably disposed towards BAS and readily agreed. Where else in the world would uninvited guests be made to feel so welcome? I was given a grand tour of the station and then visited their sauna, though not for long – the temperature was 90°C.

John Biscoe arrived four days later. After refuelling the BAS Faraday station we steamed round the north end of Graham Land to the rarely visited east coast. Here we were for the most part in uncharted waters and had to keep a close eye on the echo sounder. Turning west towards the mainland, snow was falling and at intervals we found ourselves in shoal water. *Biscoe's* path was blocked in places by the heaviest ice floes I have ever seen. Some of them had a freeboard of 1–2 metres and were covered with scattered blocks of ice as though some giant had picked up one ice floe and thrown it on top of another. If we should be trapped between two of these floes, the ship would not stand a chance.

The Master, Chris Elliott, was a glutton for adventure but now paced the bridge wondering what would come next. We were entering Prince Gustav Channel. So dense was the snowfall that no land was in sight. We could manoeuvre between the floes but more than once this led into shallow water and the command 'Full astern!' In pitch-blackness, fog and snow, we nudged into one floe, intending to stay the night. But at 0200 the next morning the floe broke up and we had to spend the rest of the night steaming in circles, dead slow. Feeling our way south, we proceeded to a point off Holluschickie Bay. This was where the geologists wanted their supplies put ashore.

Still without seeing land, we loaded the depot onto a launch and headed inshore. When the coast appeared, we followed it to a point where a small valley came down to the sea. There was a 2-metre-high erratic block of conglomerate sitting on a moraine terrace and we decided that this should be the spot. None of the people on board would be here when the geologists came, so it was important to leave everything in a conspicuous place. The bay was so shallow that we had to unload, box by box, into an inflatable boat to carry everything ashore.

The clouds lifted as we left the bay and we enjoyed a magnificent panorama of mountains on both sides of the channel. What had been a worrying 24 hours for Chris Elliott was suddenly transformed into an exciting voyage of discovery.

John Biscoe was where no ship had ever been before. Chris decided to see how far south it was possible to go. Eventually we came to a low ice front with seals basking on it. This was the northern limit of the ice shelf discovered by the Swedish explorer Otto Nordenskjöld in 1903 and first described by Alan Reece,[4] with whom I had worked 30 years before. Alan lost his life in a plane crash in Arctic Canada in 1960. From where we stood I could look north to Mount Reece, his memorial. Coincidences abound; Prince Gustav later became King of Sweden and slept through a lecture I gave in Stockholm in 1966.

At this point the ship was in 900 metres of water. We headed north, following the line of sounding from our inward track. The ice had blown across the channel towards the mainland, so for most of the way it was a clear run. At the northern entrance the Captain headed east. My diary notes:

> Found a great graveyard of tabular bergs off the Danger Islands, probably sitting on shoals. Most in 1–2 km range, probably 30 or so. Into fog at 1030, full astern for a tabular at 1200, which we just touched. Out of fog 1300, clear of all pack ice by 1400.

After resupplying Signy in the South Orkney Islands, we reached Stanley on 15 February. Giles Kershaw came on board for lunch. He was now homeward bound with Gerry Nicholson in the TransGlobe Expedition's Twin Otter, having completed the establishment of the expedition's inland station.

He offered to carry me home in exchange for nothing but a share of the flying. I was sorely tempted but decided against it on the grounds that BAS was trying to stay at arm's length from private expeditions.

That was the end of my season. I flew home via Buenos Aires, Rio de Janeiro and Madrid.

Camp on Rutford Ice Stream with the Sentinel Range behind

Unloading at sunset, Longing Gap

Tabular iceberg in the Weddell Sea

The British Antarctic Survey headquarters in Cambridge (1980)

Giles Kershaw with the Tri-turbo DC-3 at Rothera

Ian Dalziel

Garry Studd

Bryan Storey

15. WAR AND PEACE (1982)

Summoned by the Prime Minister

Without warning or provocation, armed forces of the Republic of Argentina invaded and occupied the Falkland Islands on 2 April 1982. South Georgia was attacked on 3 April. All 13 BAS personnel and 22 Royal Marines at the BAS station on King Edward Point were taken prisoner at gunpoint. The United Nations Security Council met in emergency session the same day. UN Resolution 502 required Argentina to cease all hostilities and immediately withdraw her forces. This mandatory directive was ignored. Faced with naked aggression, the British Government's reaction to the invasion was swift. Parliament voted to despatch a task force to the South Atlantic.

The United Nations and the warring parties spent the next six weeks trying to negotiate a peaceful solution but no formula was found. As far as Britain was concerned, the islands taken by force must be recovered by force.

While the main body of the British task force was steaming south, HMS *Antrim* and HMS *Plymouth* bombarded the Argentine garrison in South Georgia and secured their surrender on 25 April. A month later, on 21 May, the task force now off the Falklands landed 5,000 troops in San Carlos Bay.

These events occurred outside the Antarctic Treaty area, so had no immediate effect on BAS operations south of latitude 60°S. However, *Bransfield* and *John Biscoe*, separate but northbound at the end of the season, found themselves in a war zone. If detected by the Argentines, they could well have been attacked despite having no connection with military operations. As it turned out, both managed to escape. The British prisoners were held on board the Argentine naval transport *Bahía Paraiso* and taken by a circuitous route to the mainland. They were released to Uruguay on 20 April.[1]

BAS was plunged into contingency planning. However, there was one contingency that none of us had planned for. While I was working late in the office one evening at BAS Headquarters in Cambridge, my telephone rang. A voice identified itself as coming from No. 10 Downing Street. My first reaction was that someone could be pulling my leg. But no, it was genuine. The voice asked whether I, together with Dick Laws and Ray Adie, would be prepared to speak with Prime Minister Margaret Thatcher about the Antarctic, South Georgia and the Falkland Islands. The call evidently came to me because Dick Laws, the Director, was in Tasmania, and Ray Adie, the Deputy Director, had gone home.

With no time to ask Ray, I agreed that he and I would be happy to oblige. The voice was courteous and asked when it would be convenient. As far as I was concerned, we would come day or night, whenever the Prime Minister decreed. There was a war on – though the government insisted on the lesser term, conflict. It was agreed that we would appear at 1800 on 1 June.

BAS had no secure telephone lines, so I was reluctant to ask for details. Ray and I had to guess what questions to prepare for. In the circumstances, it was not difficult because each of us had considerable experience in British dependent territories in and around the war zone. Ray busied himself with gathering published and unpublished reports about geology and terrain, while I assembled maps and other material.

Meanwhile, the United States – officially neutral in the conflict – was feeling a need to be informed about the disputed area. One of many measures they took was to secure satellite images. NASA Landsat satellites had been orbiting the earth for ten years but the South Atlantic had been given the very lowest priority. No good pictures of the Falkland Islands were available. NASA now waited for one of the rare cloudless days over the Falkland Islands and secured an excellent set of images between 21 and 23 April.

Unconnected with the conflict, I had been using satellite images of the Antarctic for the past eight years. My contact was Dr Richard S. Williams of the US Geological Survey. Unbeknown to me, Richie learned of the existence of the new images and set about securing two copies of each for me. By pure chance, they arrived at BAS on 31 May, the day before Ray and I were to go to Downing Street. Regardless of whether or not the Ministry of Defence was aware of these, it would look good to come equipped with the latest.

Ray and I set off for London with bulging briefcases. Aiming for an inconspicuous arrival, we asked a taxi driver to put us down on the east side of Whitehall. Crossing the road, we worked our way through a crowd of onlookers to the police barrier at the Whitehall end of Downing Street. It was exactly 1800.

I informed the policeman on duty that we had an appointment at No. 10. 'Names?' he said.

'Adie and Swithinbank,' I replied.

Without a moment's hesitation, he opened the barrier just enough for us to squeeze through. We felt rather conspicuous marching down the middle of Downing Street.

To the policeman at the door of No. 10, I said, 'We have an appointment with the Prime Minister.'

'Names?' he said.

'Adie and Swithinbank.' At this point the door opened. There must have been a spy-hole.

Some staff eyed us as we were led through the hall towards the main staircase but nobody asked to examine my briefcase although a cardboard map tube resembling a bazooka was sticking out of it.

A vast oil painting of Winston Churchill dominated the bottom of the staircase. At the top we were ushered into a writing room looking rather too tidy to serve as an office. The Prime Minister, impeccably dressed, neatly coiffured, with light make-up and every last hair in place, smilingly greeted us. We introduced ourselves, American-fashion. She led the way into a comfortable drawing room on the north-west corner of the building overlooking Horse Guards Parade.

We were invited to seat ourselves on sofas on opposite sides of a coffee table. Ray faced the PM while she and I shared the other sofa. We were quickly put at ease and offered a drink. Both of us opted for gin and tonic. A private secretary hovered in the background and set off to pour the drinks. She had a glass of whisky.

The PM began: 'The reason for inviting you here is that we have been caught out in the Falklands, we have been caught out in South Georgia, and I am determined that we will not be caught out in the Antarctic. In my job, I am expected to know something of a vast range of subjects. Some I do know, but for others I have to ask the advice of experts.'

We were then launched into an intense question-and-answer session that continued without a break. What were the ramifications if the confrontation spread to the Antarctic? For how long could the British bases defend themselves if attacked? Our answer was that the bases, manned by unarmed scientists and support staff, would be bound to surrender without putting up a fight.

What was the purpose of the BAS bases in the Antarctic? How many were there? What kind of people manned them? What kind of studies were under way? Ray responded with lucid explanations. Unlike many politicians whose eyes glaze over on listening to more than a child's dose of science, Mrs Thatcher's early years as a chemist meant that it was easy to hold her attention.

Sensing an opportunity, I explained that in contrast to the Argentine and Chilean presence in the Antarctic, ours was entirely civilian. No military personnel were involved at our bases. HMS *Endurance* divided her time between the Antarctic, the Falkland Islands and the Falkland Islands Dependencies. Although scientists do need expensive instruments, they still offered the most cost-effective means of maintaining a presence in Antarctica. This paradox was explained by the military's use of far greater numbers to man a base and the soldier's wish for short-period deployments. In contrast, many BAS personnel served for two years in the Antarctic without interruption. All of them were volunteers on normal salaries; they received no incentive pay to compensate for the isolation.

In a momentary misunderstanding, I suffered looks that could kill. We were discussing Argentine claims to British Territories. At some point I said: 'Well, they do have a claim . . . ' Before I had time to continue with ' . . . but it has no validity in international law', she thought that I meant a *valid* claim. I chose my words more carefully after that.

145

In contrast to what we expected, Mrs Thatcher was a good listener, even when our answers fell short of being succinct. Eye to eye as we spoke, she was alert and extremely quick on the uptake. One question answered and then it was on to the next, without hesitation. It reminded me of the best television or radio interviewers – always ready with a new question whichever way the conversation turned.

What was the terrain like? If confrontation became a permanent feature of the South Atlantic, would it be possible to build an airfield on South Georgia, and if so where?

Ray and I grew more confident as we seemed to be providing satisfactory answers – or perhaps it was the gin. The private secretary sat in the background with notebook in hand. From time to time when the PM was particularly interested in something we said, she glanced at him with a scarcely perceptible nod, as though bidding at an auction. She meant: Write that down.

By now our glasses were empty and she offered a refill, which we accepted. As the gentleman in the background stood up and walked towards the door, she said 'Bring the bottle!' That, I felt, was a good sign.

I unfurled the satellite images, basking in their topicality as we pointed to the area where British troops were advancing towards Stanley. Ray produced BAS published reports on the geology of the Falkland Islands that related to her ideas of one day building an intercontinental airport.

I was wondering how our meeting would be terminated. Surely she had many demands on her time apart from our animated but at the same time relaxed discussion. Should I look at my watch? No, that would be tactless. We ourselves could never be bored by discussing subjects so close to our hearts, so we made no move. Eventually, she eased forward on the sofa, saying: 'On your way out, I would like to show you that I have arranged for portraits of scientists to be hung here.' The implication was that all three of us being scientists, we might be more appreciative of the pictures than mere politicians.

Walking through the Cabinet Room to the landing, the PM said: 'We have a small flat upstairs. As you go down, do look at the portraits of prime ministers on the staircase.' We shook hands and parted, she up, we down. We had been talking for 90 minutes.

Ray and I had a good laugh when it was reported that in the same week, Ronald Reagan, President of the United States, had secured a 30-minute meeting with the Prime Minister.

The Argentine forces in the Falkland Islands surrendered to a superior British force two weeks after our meeting – on 14 June 1982. In due course, the Cabinet carried out a review of the British presence in the South Atlantic. Luckily for BAS, they voted to strengthen it. BAS took advantage of this renewed appreciation of our presence in the area. With the help of NERC, Dick Laws proposed a substantial increase in our activities. It fell on fertile ground. The strict financial and staffing restraints under which we had operated for many years were eased by an increase in our net budget from £6.5 million in 1982–83 to £10.2 million in 1983–84 and £11.5 million in 1984–85. Ray and I would like to think that we played some small part at the start of this transformation.

I learned later that it was Lord Shackleton who had put forward our names to the PM as persons knowledgeable about the Antarctic. He had for many years been a staunch supporter of BAS in government circles as well as a personal friend. I also learned that our superiors disapproved of Mrs Thatcher's habit of calling in experts rather than summoning officials who might then consult experts. We should have notified NERC in advance. However, aware that they might have tried to substitute other names, we had no regrets and would do the same tomorrow. So would Margaret Thatcher.

16.   THE CHILE CONNECTION (1983–84)

Denied a parachute

My next trip south was in 1983. BAS had acquired a third Twin Otter to cope with the growing demands of a larger Earth Sciences Division. Each Antarctic spring, in October, all three aircraft flew south by way of the west coast of South America. On 19 October they departed southbound from Lima, Peru, on flight plans to Antofagasta in Chile, each one properly cleared through Lima Air Traffic Control. One hour into the flight they were ordered to land at the coastal town of Pisco. One of the three aircraft was intercepted by two jet fighters of the Peruvian Air Force, which then led it to Pisco. All our aircraft landed safely, and were then searched while the aircrews were kept at a distance by armed guards. On one of the aircraft, even fixed panels were unscrewed to see what might be behind.

The aircraft and crews were held for 22 hours. Finally they were told that no regulations had been broken and they could proceed. One possible reason for all this hassle was that Peru, which was on the side of Argentina in the Falklands war, wished to show its displeasure by flexing its muscles. No official explanation was ever forthcoming.

On arrival at Antofagasta, some sugar was found in the filler neck of a fuel tank in the aircraft that had attracted the most interest the day before. Only Peruvians could have had access to the aircraft while it was detained overnight in Pisco. In plain English, it was attempted murder. However, the on-board fuel filtering system did its job.[1] We never found out whether it was an official act, or merely the act of a disgruntled individual.

Leaving Cambridge two weeks after these events, I flew south via Santiago, Chile. I had earlier arranged to collaborate with Fuerza Aérea de Chile (FACH) on an Antarctic mapping project. In Santiago I was led by Cortland Fransella, Counsellor at the British Embassy, to meet General Mario Lopez. Lopez had very kindly arranged for FACH to assist in purging the spiked fuel of our Twin Otter, for which I thanked him profusely. He had little interest in the mapping project but would arrange a meeting with 'the only man who can get things done'.

That was General Javier Lopetegui Torres, Adviser on Antarctic Policy to the Chief of the Air Staff. He was a former pilot and now an Antarctic specialist with wide interests.[2] The next morning I found Lopetegui and was introduced to others who might be involved in mapping. We were cordially received at Servicio Aérofotogrametrico at Los Cerillos airport, after which the Commandant showed us over their facilities. Their equipment and map output were both of excellent quality, and on leaving, I was presented with a series of aeronautical charts extending from Peru to Cape Horn – a distance of more than 4,000 kilometres. These were constructed according to international standards for the World Aeronautical Chart series – with one important exception. I found that not a single airport was shown anywhere in the adjoining parts of Peru, Bolivia or Argentina, although there are many.

I pity the poor pilot who strays across the border and then needs to land. This sad omission was a measure of the almost permanent state of tension that exists because of border disputes.

I flew on to Punta Arenas to await the arrival of *John Biscoe*. She had already made one trip south and was returning to Punta Arenas to pick up passengers. In the event, she was delayed by one of the roughest crossings of Drake Passage on record.

At dinner in the hotel Cabo de Hornos on 14 November in the company of a large contingent of Fids, I sat facing east towards the Strait of Magellan. A strange hybrid aircraft flew past the picture window – it resembled a DC-3 but with a pointed nose. Those of us who recognized it knew at once that – of all the aviators in the world – only Giles Kershaw could be flying a machine like that. Who else would be giving his passengers a low-level, close-up view of the city of Punta Arenas?

Giles had brought the Tri-Turbo, a DC-3 fitted with three engines instead of two. This aircraft was unique in the true sense – only one was ever built. It was modified from a standard DC-3 by replacing its relatively unreliable piston engines with modern turbine power plants of approximately the same horsepower. So impressed was Giles with the performance of this half-breed that he had chartered it from the owners, Polair of California, to fly a group of climbers to Vinson Massif in the Ellsworth Mountains – the highest mountain in Antarctica.

Once again Giles was defying the polar establishment and its wish to keep Antarctica exclusively for government-sponsored work in support of science.

The Chilean Government, however, had agreed to support the expedition. They saw that if Chilean facilities on the mainland and in Antarctica were to offer the key to increasing activities in the Antarctic, Chile stood to gain economically as well as strategically. The strategic argument was that, whatever the future political status of Antarctica might be, nobody would seek to exclude the state that held the key to easy access.

Giles' party included nine people calling themselves the Seven Summits Yuichiro Miura Antarctic Expedition. Lopetegui and I joined them later in the bar of the hotel Los Navigantes. One was Chris Bonington, who introduced me to 54-year-old Frank Wells, ex-President of Warner Brothers and a 25 per cent sponsor of the expedition; 53-year-old Dick Bass, a Texas oilman, owner of the Snowbird ski resort in Utah, and also a 25 per cent sponsor; Yuichiro Miura, another man in his fifties, who had skied down Mount Everest and was a 50 per cent sponsor; Rick Ridgeway, climber; and Beverly Johnson, also a climber but here in Chile to do some filming.

These were *la crème de la crème* of the world's mountaineers, some of them with the money to pursue their esoteric ambitions. Wells and Bass were trying to climb the highest summit on each of the seven continents within the space of twelve months. They had already done Elbrus in the Caucasus, McKinley in Alaska, and Aconcagua in Chile. They had been forced to give up at 8,200 metres on Everest. Now, before the end of 1983, they proposed to climb Vinson in Antarctica, Kilimanjaro in Tanzania, and Kosciusko in Australia.[3]

The next morning Lopetegui invited me to fly with him to the Antarctic in exchange for some advice on possible sites for a new FACH base. I knew that helping Chile in this way would be very much in the spirit of the Antarctic Treaty, so I was inclined to accept. There was another, even better, reason. Much to BAS's embarrassment, a FACH contingent had been living, uninvited, on our snow runway at Rothera. They were nice people but their camp was causing potentially dangerous snowdrifts in the aircraft taxiing area. If I could find a site for them to move to, a potential source of friction with BAS would be averted and everyone at Rothera would be grateful.

I was woken at 0400 on 17 November, my fifty-seventh birthday, and taken with Lopetegui to the airport. We were to fly in a FACH Lockheed Hercules C-130 carrying 48 drums of avtur to supply the Chilean camp at Rothera. Frank Wells had paid $16,800 for 41 of the drums so that the Tri-Turbo could refuel there, both southbound and northbound. The drums were mounted on pallets to be airdropped with large parachutes. In addition, there were about 30 military personnel, most looking thoroughly dejected. Only ten of us could be seated; the rest were sprawled over parachutes and heaps of cargo.

After shivering in the cold-soaked cabin on the ground for one and a half hours, we taxied out for take-off. Nobody who had a seat belt bothered to fasten it. Half an hour into the flight, the cabin became unbearably hot. When eventually the rear cargo ramp was lowered for the airdrop at Rothera, we became very cold again. The pilot invited me up front to point out the sights as we did lazy drop runs over the airstrip. Not all of the parachutes opened; presumably the 20 Chileans standing in the drop zone were used to sprinting for their lives to avoid a one-tonne pallet approaching them at its terminal velocity. At this point we were only 300 metres vertically above where I wanted to be, but nobody offered me a parachute.

After completing the airdrop, the big machine headed north-north-east, straight for the new hard airstrip at the Chilean Teniente Rodolfo Marsh Martín station on King George Island.[4] We knew it as Marsh. Landing with a bump, the aircraft splashed through snow-melt puddles as it slowed after its six-hour flight.

I was ushered into the 'hotel' and given a glass of *pisco* sour, the national firewater. Things now looked rosier. The hotel is a 200-bed accommodation block used for FACH personnel and also for visitors of all kinds – even sometimes paying tourists. Dinner for the visitors was in the Commandant's private quarters – with waiters, wine and good food. I was left in no doubt that Chilean officers who volunteer for Antarctic service are thoroughly pampered. I am less sure about the other ranks.

The next day Lopetegui and I, with a dozen airmen, headed back to Rothera in two FACH Twin Otters flying in formation, each of them equipped with combined ski-wheel landing gear like BAS's own aircraft.

Landing at Rothera on a fine calm evening at 2300, I saw a handsome FACH Bell 212 twin-engined helicopter near the BAS Twin Otters and suggested that, night or day, good weather like this should not be wasted. We should, then and there, go on a tour of possible base sites. Summoned from their well-oiled party – a factor I had failed to consider – the pilots rolled into the helicopter and off we went. Comandante de Grupo Dario Bobadilla, Lopetegui and I were in the cabin. Venturing east into the fjords along the Graham Land coast, we were shaken up by severe turbulence and the officers decided that they did not fancy the idea of living there.

We landed back at the Chilean camp after midnight. The Rothera skiway, where the Chileans had established themselves, was 3 kilometres from Rothera station. Snowdrifts had deeply buried their officers' quarters, which we reached down a set of snow steps. Eight of us slept in a bunkroom with an adjoining kitchen and dining table. The officers seemed to be living comfortably as troglodytes – the other ranks made do with tents in the open.

In the morning we flew in the helicopter to visit BAS's old Adelaide station on the south coast of Adelaide Island, landing on the main street. The place had been unoccupied since BAS moved to Rothera in 1977. The Chileans were visibly impressed with the general good condition of the base and enquired about using it. I said that would be a matter for government-to-government discussion. To help them make a fair judgment, I listed the drawbacks which had led BAS to abandon the place: an icy uphill runway with a side-slope, an ice ramp at times impossible to negotiate, crevasses inhibiting access to possible landing sites higher up, limited bare ground on which to build, and an ocean swell that sometimes made it dangerous to unload ships.[5]

Back at Rothera, I took leave of my hosts on being delivered by helicopter to the front door of the BAS station. It had been a delightful interlude. I have always found Chileans to be generous people and easy to work with. Now I had been able to help them in return.

Lopetegui reappeared some days later to ask if I would agree to fly with their Twin Otter to search for a site for a fuel depot. They were hoping to extend their operations southward but seemed to have little feel for terrain and were nervous about venturing into new areas. I was happy to help.

I had advised them to put the depot on rock so that the drums would not be buried under snowdrifts. To achieve this, they wanted a site about 500 kilometres south of Rothera where it was safe to taxi right up to a nunatak. After discussing various options with John Hall, the Rothera base commander, I suggested we should try Coal Nunatak on Alexander Island.

The Captain of the Twin Otter was Victor Rodriguez; Lopetegui also brought five airmen for training. Three hours later we landed off Coal Nunatak, but without any reconnoitring of the area for crevasses. Without intercom, I had been unable to caution them. Taxiing to a rock platform, we walked about for some minutes while the crew pumped two drums of fuel into the main tanks. While my attention was diverted, three of the passengers marched off towards another outcrop, blissfully unaware of some nearby crevasses. It was embarrassing for me, their guest, to have to bellow at them to stop.

Lopetegui took advantage of the moment for some brief but formal words of thanks to me for helping them to find a good site, after which we sped home with a tailwind. Over Marguerite Bay, the plane was forced down by cloud to 100 metres over the sea. Sitting as I was, right behind the pilots, I could see that they were having an animated discussion about which way to head for Rothera. Eventually the Captain turned to me and asked me to point the way. It reminded me of the time, many years earlier, when I was involved in minesweeping in the North Sea. Norwegian fishing boats used to hail us with: 'Which way to Aberdeen?' They seemed grateful when we pointed with one arm.

At Rothera we heard over the radio that Frank Wells had asked NSF to allow him to purchase 14 drums of fuel from Siple station, so that the Seven Summits expedition could fly to the South Pole, and another 22 drums from the South Pole to provide for their return to Rothera. As a businessman, he expected that money would open doors. But NSF is not in business, so the answer was no. Chile's contrasting attitude was made clear when Lopetegui offered to sell Wells as much fuel as he wanted. In the event, the expedition was unable to accept the offer.

On 7 December the Seven Summits expedition landed once again at Rothera, having completed their climb of Vinson Massif. Frank Wells had a badly frostbitten nose to show for it. Four hours later they continued to Punta Arenas – with Lopetegui as their guest.

Lopetegui had a real interest in the Tri-Turbo because converting another Douglas DC-3 (or C-47) could be done in Chile. 'The only thing we have in Chile is cheap labour,' he said. Another reason for interest was that no amount of money could buy him the alternative, a Lockheed LC-130, the ski-wheel version of the Hercules. The Pentagon's refusal of an export licence was ostensibly linked to Chile's human rights record under the military junta of General Pinochet. Quite apart from that, the Pentagon, for its own reasons, has never allowed the export of ski-equipped LC-130 aircraft to any country.

I was now back in the BAS fold. For some 40 years up to 1983, our geologists had been virtually confined to the Antarctic Peninsula area. Now, with a permanent staff of professional geologists, this limitation had become irksome. The principal constraint had been logistic. The Twin Otters could venture far inland, but only if there was some means of refuelling in the interior. Flying fuel from Rothera was a case of the law of diminishing returns: it took 10–20 drums of fuel to fly one drum to an inland destination.

One of the burning ambitions of geologists on both sides of the Atlantic was to seek, in the transition zone between the Pacific coast of Antarctica and the Transantarctic Mountains, clues to the history of the break-up of the former supercontinent – Gondwanaland.

The best solution would be to mount a joint operation with American geologists. To achieve this we needed a vigorous protagonist on each side of the Atlantic. Ian Dalziel, a geologist of Scottish origin on the staff of Lamont-Doherty Geological Observatory of Columbia University in New York, was keen to work with us for his own reasons. He knew that American field camps, like the Ellsworth Mountains camp that Garry Studd and I had visited in December 1979, had helicopters but lacked the Twin Otter's ability to operate over a much greater radius from any fuel depot. For reasons best known to themselves, NSF had resisted calls from their own scientists to purchase a couple of Twin Otters.

Ian had to overcome opposition within NSF and also from certain Congressmen. The US Antarctic Program's traditional openness to foreign participants had come under fire on the grounds that it amounted to 'giving away taxpayers' money'. This was understandable, in that there had been very little quid pro quo over the years. Foreign scientists had not contributed to the tremendous logistic costs.

We were proposing something quite different – an equal partnership in logistics as well as in science. Each side would contribute what they did best. In exchange for fuel put in by NSF's giant LC-130 aircraft flying from McMurdo, BAS would provide the services of two Twin Otters for six weeks. One aircraft would carry out aeromagnetic and radio-echo sounding surveys along agreed flight lines. The other would provide close field support – essentially an air taxi service – for a geological party of eight, consisting of four from BAS and four Americans.

Ian and I had to face muted opposition even from some geologists. How could we ensure that one side might not independently rush into print with the major findings in an attempt to skim the scientific cream from the joint project? I discussed this sensitive issue and we agreed to be vigilant.

Finally we overcame the obstacles. BAS and NSF agreed on a two-year joint West Antarctic Tectonics Project, and 1983–84 was to be the first season. Our geologists were at Rothera and Dalziel's were at McMurdo. Hercules flights from McMurdo had distributed 100,000 litres of fuel between Welcome Nunatak (79°S), Martin Hills (82°S) and Thiel Mountains (85°S). Garry Studd had flown in two GAs, Ian Lovegrove and Paul Wood, to be on hand for the arrival of the American party at Thiel Mountains.

Garry was now ready to fly in our geologists, Bob Pankhurst and Bryan Storey, and invited me to come with him. I was anxious to give the project a fair wind, so it was an opportunity not to be missed. On 9 December we loaded up and climbed aboard Twin Otter VP-FAZ, known by its call sign Foxtrot Alpha Zulu. This aircraft had a checkered history. Two years earlier it had been flipped on its back at Rothera by hurricane winds. The wings had to be taken off and the machine was lowered into the hold of RRS *Bransfield* for the voyage home. After being rebuilt at a substantial cost it was returned to its element. However, it was the same aircraft that had suffered an unwelcome dose of sugar at Pisco airport seven weeks earlier.

Take-off checks revealed a misbehaving fuel flow gauge. Engineer Andy Carter opened up the fuel filters, and found grains of sugar in one and grains of sand in the other. It took him several hours of painstaking work with bare hands to make everything safe.

Finally we were airborne at 1743 and landed at Fossil Bluff to refuel. The occupants of the hut fed us a good meal and we took off again southbound. Cloud forced Garry up to 3,700 metres but we were able to navigate with the help of his new colour radar, which showed up the mountain ranges ahead. It took four hours to reach Welcome Nunatak, site of the former American camp where Garry and I had roused the sleeping inhabitants in 1979. We found one of the American Jamesway huts with only the peak of its roof showing flush with the snow. It had taken only four years of snow accumulation to bury it. Digging out the entrance, we slid down a chute to get inside. After curried meat bar for supper, each of us spread a sleeping bag on the plywood floor and climbed in. Because summer temperatures had not yet reached this level, the night was much colder than it would have been outside.

In the aircraft's survival gear, Garry had a very modern short-wave transmitter, no bigger than a domestic transistor set. In the morning, with only a wire aerial led up to a post outside, and without leaving his sleeping bag, Garry spoke with Rothera (now 1,400 kilometres away) and got a weather report. Conditions were poor at the Thiel Mountains but might be good enough for us to reach Martin Hills 350 kilometres away. So after breakfast we took off and headed south.

The Martin Hills depot was now more conspicuous, Garry reported, than when he saw it two weeks before. There were four large cargo pallets in a row. In order to save many hours of Twin Otter flying time, the Americans had agreed to provide eight skidoos, eight sledges, a Jamesway hut with the fuel to heat it, and all the skidoo fuel. A giant fuel bladder had been brought in the previous summer. BAS was providing tents and field rations for both the US and BAS contingents.

On landing at Martin Hills, we unloaded the cargo and returned to Welcome Nunatak for another load – this time of man food. The American party had been put down 900 kilometres further south at the Thiel Mountains on 1 December, the date that Ian and I had agreed on seven months before. BAS had technically defaulted by not delivering our geologists and a Twin Otter at the same time, so we were anxious to join the American party to start the fieldwork. The weather was marginal and we flew on top of solid cloud, hoping for an opening through which to let down on reaching the Thiel Mountains.

By dead reckoning, we were only 10 kilometres from our destination when a hole appeared. Lovegrove and Wood had spent several days wielding shovels to prepare a 'runway'. This involved slicing off the tops of all sastrugi and shovelling the debris aside. The finished runway was 300 metres long and barely 10 metres wide. Garry had to do some fancy footwork to keep from straying off it. However, the great joy of a runway of this kind is that pilots can stray off it without courting disaster – just a bumpier ride.

The American party consisted of three geologists: Ian Dalziel, Walt Vennum and Anne Grunow, together with Chuck Kroger, a GA. When we newcomers had made camp, the combined party numbered ten people housed in five pyramid tents. I went to Ian and Anne's tent for supper and a discussion on plans.

The geologists' first target was the Thiel Mountains, a jagged, snow-capped mountain wall that was spread out before us. For me it was an evocative moment because I had been working with Edward Thiel when he was killed in a plane crash 22 years before. These mountains were a fitting memorial – how wonderful that they would bear his name for ever.

Garry and I remained in camp to tidy up, and later started back for Martin Hills to fetch some drums of fuel. We had no weather report because there was nobody there to report it. After flying north for an hour, we saw that the cloud layer below us extended right down to ground level. If we returned to Thiel Mountains, we would have to burn an extra three drums of precious fuel, when the weather improved, to get back to where we were now. Garry did the sensible thing. On finding the nearest place where he could see the snow surface, he asked me to count down his radio altimeter reading while he concentrated on the flying. Seconds after I read out 'Twenty feet . . . ten . . . five feet' there was a bump; we were on the ground. We pitched camp, secure in the knowledge that the aircraft carried a vast amount of survival gear. Many pilots I have known would have chosen to return to their starting point, regardless of the waste of fuel.

The next morning we took off in sunshine, collected our load from Martin Hills and flew back to the Thiel Mountains. Now Garry's air taxi came into its own. I stayed in camp while he took the geologists for a day trip to Stewart Hills, 100 kilometres away. They returned in time for supper, euphoric because of the ease and efficiency of travelling in this fashion.

Rothera station (February 1986)

Ice floes off Hampton Glacier

Camp in the Thiel Mountains

Garry Studd taking off from the 'hand-made' runway at the Thiel Mountains

Flight to the South Pole

Welcome at the South Pole (© Anne Grunow)

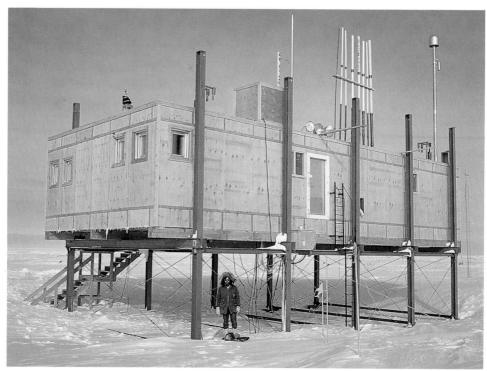

The 'Clean Air' building at the South Pole

Passing Shackleton Range on the flight from South Pole to Halley

17. THE SOUTH POLE (1983–84 continued)

A royal welcome at 90 degrees south

It now became clear that we would need extra fuel for the number of Twin Otter excursions the geologists had in mind. Another Hercules flight from McMurdo, even if NSF agreed to it, might take some time to materialize. Noting that we were only 560 kilometres from the American South Pole station, Ian enquired of the NSF representative by radio whether we might fly to 90°S a few times to refuel. This was readily agreed. Another reason for doing this was that all the rocks we had collected could be put on any one of the many Hercules flights returning empty to McMurdo.

It was midnight – but the weather was fine. Although 19 hours had elapsed since Garry's first take-off that morning, he said, 'Let's go!' As ever, it went without saying that good weather is for work, bad weather is for lolling in one's sleeping bag with a good book. We loaded six empty fuel drums into the cabin and headed due south. Ian came along to ensure that everything went smoothly at our destination.

I flew most of the way, while Garry looked after the navigation. Heading off on a magnetic compass bearing with only one thing between us and the Indian Ocean gave pause for thought. With no landmarks, it was vital that we should not stray off track. A pilot would have a very red face if he ever called the Americans to say that he was out of fuel and uncertain of his position. Garry was taking no chances. Using a sun compass mounted on the window sill, he shot a sun azimuth every few minutes to keep us on a true south heading. When the great shining dome of the station did appear after two hours, it was right on the nose.

Knowing that some of the Hercules pilots – without any justification – treat Twin Otter pilots as 'cowboys', Garry felt it would do no harm to demonstrate a bit of professionalism. He asked the Air Traffic Controller for a practice ground-controlled approach to the runway. This was agreed. The sun was shining but Garry looked down and glued his eyes to the instrument panel. The controller too was professional, and with a relaxed drawl directed Garry onto the glide path. I was looking out but Garry did not look up until the controller told him to. We were properly aligned and only a few metres off the ground. I hope they were impressed.

The runway is 2,800 metres above sea level and 4,270 metres in length, perhaps the longest and smoothest in Antarctica. As we taxied up to the station it became evident that our arrival was something of an event for the locals. Around 40 people were gathered outside the main entrance, ogling what was to them a very small aircraft. Some were waving, others photographing. I opened the front window to acknowledge the friendly reception. Welcomed like royalty, I felt bound to wave and nod like royalty.

Wondering where to go for fuel, we saw ahead a giant newly painted BP sign, with its prominent yellow lettering on a green shield background. They had done it all for us. It was a joke, yet at the same time a measure of their welcome, and I found it very moving. We filled the aircraft tanks and the six drums that we had brought with us. Bob Hurtig, the Station Manager, invited us in for a three-course meal. A ramp led down to a snow-floored tunnel. Over the entrance hung a sign declaring: THE UNITED STATES WELCOMES YOU TO THE SOUTH POLE. The tunnel led into a giant aluminium geodesic dome, 50 metres in diameter and 17 metres high. Under the dome there were three two-storey prefabricated huts. One housed laboratories and sleeping quarters, another contained the kitchen, dining and recreation rooms, and a third was for radio communications and a library. A series of steel arches housed the garage, gymnasium, carpenter's shop, power plant, hospital and fuel store. Giant fuel bladders laid straight on the snow under one of the arches could hold 950,000 litres. The fuel was both for the station and for aircraft. The amount was such that, if they gave it away, our Twin Otter could fly for 3,000 hours without running out.

The proper name of the place is Amundsen-Scott South Pole station and it is the showpiece of the US Antarctic Program. About 80 people were in residence at this time of year. They were engaged in research in meteorology, air chemistry, seismology, the upper-atmosphere, aurora, cosmic rays and astrophysics.

In summer, the South Pole is the sunniest spot on earth and often has 24-hour days of uninterrupted sunshine. Day and night are indistinguishable. The sun circles the sky at the same height above the horizon. Decisions about when to work and when to sleep are arbitrary. For convenience, the station keeps McMurdo time, which is also New Zealand time. Taken literally, a stay 'overnight' could imply wintering here, because each night lasts six months – from 21 March to 21 September.

The air temperature was –31°C and there was no wind. The annual mean temperature, I learned, was –49°C; the lowest ever recorded was –83°C, the highest –14°C. Because I was well-dressed – in the Antarctic sense – the cold was no problem, but suddenly arriving at this altitude led to heavy breathing after exertion.

The station was carefully sited about 350 metres upstream from the geographic South Pole – towards which the ice (and with it the station) moves at a rate of about 10 metres per year. Once beyond the pole, the ice will be moving along the meridian of 40 degrees west of Greenwich towards Filchner Ice Shelf and the Weddell Sea.

Before leaving, I took the opportunity to mail a bundle of letters from the Thiel Mountains crew. After three hours on the ground, surrounded by what seemed like an army of people attentive to our every need, we were airborne not only with all tanks full but also with the six cabin drums replenished, plus a carton of American ice cream to cheer the geologists. Thanking everyone for their hospitality, we headed back towards our six little tents in the wilderness. When we landed at 0730, Garry had been on the go without a break for 26½ hours. Glowing with satisfaction, he said it was the greatest day of his life.

Privately, I had known for years of his ambition to fly to the South Pole. More than once I had said: 'Hang on, we'll get you there.' Now at last I was off the hook. It was his seventh and last season – he had hung on even longer than Bert Conchie, who had served BAS from 1969 to 1975.

The Thiel Mountains camp was the most southerly in BAS history and it took everyone some time to come to terms with the weather. The following days demonstrated that geologizing and travelling at –20°C with a 20-knot wind blowing fine snow into your face was more the rule than the exception. There were lie-up days when the wind was too strong but on each flyable day the geologists clambered into the Twin Otter for excursions, while I stayed in camp to man the radio.

Our second South Pole flight was on 18 December. The geologists had spent the morning at the southern extremity of the Thiel Mountains and the afternoon at the northern end, returning at 1830 for supper. Garry and I took off for the South Pole at 2000 with Anne Grunow as passenger. We timed these fuel runs to be at night – by BAS time – so that we arrived during day-time – South Pole time. This was convenient for everyone except Garry, who had to sleep sometime. This time it was only the Station Manager and his ever-attentive refuelling crew who were outside to welcome us.

The station doctor, who doubled as shopkeeper, kindly opened his shop so that we could post letters and buy souvenirs. From there we went to dinner, which we gladly accepted without admitting that, according to our circadian rhythms, it was breakfast. Having found on our last visit that, in spite of the high altitude, we could get airborne with six full drums in the cabin, this time we took seven. The plane used more of the runway but still came off the ground. On the return flight, Garry gave Anne her first flying lesson while I dozed in the back.

The following day Garry and I slept the clock round; we needed it. But the pattern of taking the geologists where they wanted to go was highly successful. One day the British contingent drove skidoos to the nearest outcrops while I flew with the Americans to Moulton Escarpment. By 21 December both groups had completed everything they wanted to do within range of the Thiel Mountains camp and fuel supply. They were tired, delighted with the ease and versatility of Twin Otter geology, and had a tonne of rocks to show for it.

The plan was now to move camp north to Martin Hills. It took three flights to do it, each round trip taking 2 hours and 45 minutes. Once Garry let me handle the landing, but I touched down clumsily before the start of the runway. Ian, who witnessed the event, asked Garry why he had landed short. I felt honour bound to reply: '*He* didn't!' Ian understood.

Chuck Kroger, a carpenter by trade, and Walt Vennum had spent the day assembling the 5 by 5 metre Jamesway hut brought from McMurdo in one of the Hercules flights. Although BAS had agreed to provide the food, my diary records that I walked into the hut to find 'a great variety of expensive and unnecessary complications'. With the benefit of hindsight, that was a harsh judgment from one who was about to tuck into American lobster tails, fillet steak and a variety of other foods that by BAS standards were pure luxury.

As Christmas Eve brought poor weather for flying, we decided to use the day for feasting – Scandinavian style. Garry assumed the chef's position, helped by Ian and Anne. The menu was distinctly geological:

Prawns à la Thiel

Sopa de la Nash

Florida Lobster tails granite

Texas steak porphyry

Christmas pudding with brandy butter

Cheese and biscuits

Coffee, beer, three kinds of whisky

Three table wines and four liqueurs

It was not just the alcohol that made me feel that our joint operation had been a greater success so far than I had dared to hope. Here we were, ten people from quite different backgrounds, working for a common cause and agreeing without difficulty on how the job should be done. We were equal partners – as Ian and I had intended. Part of the explanation is that both parties were older and more experienced than many of the people who come to the Antarctic. I had insisted on this because of the unusually high cost of the operation for NSF as well as for BAS. Moreover, we were to collaborate on the project for a second year, so continuing goodwill was essential.[1]

Hangovers notwithstanding, Boxing Day saw the geologists 250 kilometres away on a visit to the Whitmore Mountains and Mount Seelig. Garry told me that his first landing was on rough sastrugi, so while the geologists were doing their work he used a shovel to smooth a runway for take-off. Where else in the world, I wondered, could a pilot construct a runway without assistance – between breakfast and lunch?

Preparations for the following season were already under way. On 28 December we spoke with a Hercules flying overhead to put down a fuel depot in the Jones Mountains. Another Hercules topped it up the same day.

The time had come for Garry and me to leave. It had earlier been decided that, in the best interests of the project, Garry's six-week contribution should be divided into two three-week periods. During the interval, the geologists would use skidoos to reach nunataks in the vicinity, completing their studies at a more leisurely pace than was practical with the Twin Otter. BAS needed the aircraft to ferry Dick Laws and others from Halley to Rothera.

We bade farewell to our colleagues and headed south. Landing at the Thiel Mountains, Garry and I loaded about a tonne of rocks to be carried home via McMurdo and flew on to the South Pole. This dovetailed nicely with our need to reach Halley. There was no fuel left at Thiel Mountains and the runway was too bumpy for taking off with a heavy load. We would ask the Americans to fill us up for the flight to Halley.

Landing at the pole, we unloaded the rocks and Garry helped the refuelling crew to fill everything to the brim, including every drum in the cabin. Meanwhile I went to post letters. Opening the post office, the doctor found that he had run out of stamps. His pragmatic solution was to cut stamps from philatelic mail and hand them to me for sticking onto our letters.

Philatelic mail is the bane of an Antarctic postman's life. Collectors from all over the world send thousands of ready-stamped, empty envelopes to be returned to them. They want the station cachet and some signatures to authenticate it. Many of the senders are dealers who send dozens of envelopes for franking. Handling the flood of mail can take many thankless hours.

Some of the philatelic mail arrives in envelopes addressed to an individual. My own practice is to return one franked envelope to each of the senders for their own collection. I feel under no obligation to handle bulk mail for the benefit of someone else's business. As for the rest, I and my colleagues use the South Pole doctor's solution.

Garry and I needed a nap and we were offered bunks in the sleeping quarters. This was something of a privilege because most summer visitors are assigned to a camp bed in a Jamesway hut outside the dome. Our room had two bunks, one above the other. It was cosy but, having no windows, also slightly claustrophobic. Oxygen breathing sets hanging in the corridor showed that the residents were well aware of the fire hazard.

We tried to sleep but, since it was daytime for everyone else, our need for sleep was not widely known. We were continually interrupted by the public address system calling, 'Pilot of VP-FAZ to radio shack please,' then 'Twin Otter Garry to comm centre please.' A short time later, 'Alpha Zulu pilot to pick up incoming mail please.' Five LC-130 flights arrived – all too audible even in here. I must have had some sleep before Garry woke me to say that the weather at Halley was good.

We both had a shower. There was a unisex bathroom to underline the change from misogyny to equality of the sexes. Men and women were trusted not to peep behind the shower curtain to see who was inside.

Thanking everyone for their ever-cheerful help, we took off at 0750. Garry assured me that we were heading true north. Since northward lay the Atlantic, Pacific and Indian oceans, I was not sure how to respond. But the sun was dead ahead of us – which is where it was supposed to be. The glare in our eyes was quite exasperating.

Garry had been so anxious to get going that he had omitted to respond to the call of nature. Handing me the controls so that he could go aft, he said, 'What will you do if an engine stops?' I said I would trim up and make do with the other. 'Well, don't! Just chop the power and glide at a hundred knots.' He was of course right not to trust me with anything more advanced.

He had chosen to take us along the meridian of 30°W, although a direct route to Halley would be along 28°W. The logic was that if he followed 28°W and eventually found the coast under cloud, he might not know which way to turn to look for Halley. By keeping to 30°W he knew that it could only be a right turn to Halley. It was a sensible precaution.

We took turns flying, changing every hour. After two and a half hours I spotted the Pensacola Mountains 300 kilometres to our left, so we were on the right track. The distance we had to cover was 1,600 kilometres, so somewhere along the way we would have to land to pump in fuel from the drums in the cabin. Four hours out of South Pole, we spotted the Whichaway Nunataks right ahead. I bet Garry half a case of champagne that he could not find the remains of South Ice, the tiny outpost of the Trans-Antarctic Expedition of 1955–58. We knew that in spite of the interval of 27 years since it was occupied, it had been sighted not long before by Russian airmen. Although we must have passed within 1 kilometre of the place, we saw nothing.

A few minutes afterwards, Garry landed among the Whichaway Nunataks. Again, this was a sensible precaution because if, for any reason, we failed to take off again, we could tell the other BAS aircraft where to look for us. On the ground, it was calm and very much warmer than where we had started from. We felt even warmer after the effort of pumping four drums of fuel into the tanks.

Starting up again, there was a panel light showing low oil pressure on the left engine. This was no place, I thought, to have an engine problem. Garry must have felt the same, because he simply disconnected the warning light. Now with no warning light on the instrument panel, we took off. Pilots are a pragmatic breed.

We enjoyed a magnificent view of Recovery Glacier sweeping across our path towards Filchner Ice Shelf. Then Shackleton Range, the great Slessor Glacier and Theron Mountains. We found Halley without difficulty and circled RRS *Bransfield* before landing at the station. Dick Laws, Eric Salmon and Paul Whiteman were there to greet us.

164

There were two Soviet ice-strengthened cargo ships secured against fast ice not far from where *Bransfield* was unloading the year's supplies for Halley. Seven months before, I had written to a Russian colleague asking whether Soviet Antarctic Expedition ships, westbound along the Weddell Sea coast, would consider carrying 100 BAS fuel drums from Halley to one of their summer stations: Druzhnaya I on Filchner Ice Shelf or Druzhnaya II on Ronne Ice Shelf. This would provide a fuel depot for our aircraft and also save *Bransfield* a time-consuming diversion from her itinerary. BAS offered to pay for the fuel. My contact in Leningrad was Garrik Grikurov, whom I had known, mostly by correspondence, for 20 years. It is worth quoting from his reply, because it was another shining example of international collaboration:

> Our ships will try to collect 100 drums at Halley . . . If for some reason loading operations at Halley appear impossible, your air-craft can always refuel from our stock at either of Druzhnaya bases . . . This, of course, will not involve any financial obligations on your side – we can easily spare 100 drums.[2]

Kapitan Gotsky was busy unloading two giant Mi-8 helicopters. *Pioner Estonia* was carrying two Ilyushin-14 aircraft for summer operations. I went to talk to Captain John Cole in *Bransfield*, and on entering his cabin found a group of Russians holding glasses of gin and tonic. Among them were Captain Tschelovanov of *Gotsky* and their Expedition Leader, Vladimir Ivanov, a geologist. Some drinks later, Tschelovanov said that, if we want-ed, he would take 150 drums instead of 100. I did want 150 but had not dared to ask.

Loading was completed on New Year's Eve and several of us were invit-ed aboard *Gotsky*. Hoisted in a sling from boats, we were invited into a stan-dard 6-metre cargo container lashed down on a hatch cover. Inside it lived six of the scientific staff in cheerful squalor. I felt that such conditions should be compulsory viewing for Fids, who might then realize what luxu-ry they enjoyed in our own ships in comfortable cabins below deck.

The New Year revels were well under way, and after a few drinks I felt a hand taking my long-cherished but well-worn Russian fur hat. The thief left the container. To complain about a solitary pickpocket might lead to a diplo-matic incident, so I kept quiet. Some time later we found ourselves in the Captain's cabin eating ice cream liberally doused with Georgian champagne. The whole crew went to the bridge at midnight to drink vodka toasts to peace and friendship, finally singing *Auld Lang Syne*. As we turned to leave, the pickpocket silently thrust into my arms a brand-new Russian fur hat. It was characteristic of these people.

Back at Halley, we found another New Year celebration under way. It was midnight by BAS time, some hours after the Russians had rung in the New

Year according to Greenwich Mean Time. If goodwill can be measured, I thought, surely our cup runneth over.

Garry had light-heartedly agreed, months before, to carry Dick Laws and three others on 'the 0840 Halley–Rothera flight on 2 January 1984, weather permitting'. Taking off from Halley only an hour behind 'schedule', we had a good laugh.

On the way down the coast we passed *Gotsky* and *Estonia* working ice 100 kilometres short of Druzhnaya I. Landing at the station to refuel, we were fortunate to find some full drums that BAS had put there some years before. We were invited into a cargo container for smoked salmon, salami, bread and butter, chocolate and de-icing fluid. Russian de-icing fluid is 96 per cent pure alcohol, without additives to discourage ingestion. Except for Garry, all of us drank some.

A Russian helicopter pilot gave me his Aeroflot cap badge and drove us on a sledge back to our aircraft. From here we flew the 1,500 kilometres to Rothera, refuelling at Fossil Bluff on the way.

At Rothera I was somewhat embarrassed to find an invitation from the Operations Directorate of FACH to fly in their C-130 aircraft from Marsh station to Punta Arenas on 16 January. The invitation was addressed to me personally, whereas my boss, Dick Laws, had a greater need than I to fly home as soon as possible. The matter was diplomatically solved by arranging for the invitation to be extended to include Dick and Eric Salmon.

We spent the next few days visiting geologists and glaciologists in the field. Garry's aircraft, with its cargo of senior staff 30 years older than most of the men we were visiting, was now known as the geriatrics' air taxi. One day we landed on a frozen sea lake at Ablation Point near Fossil Bluff. An easterly wind forced Garry to make a steep turn to align the aircraft with the best place to land. At a 60-degree bank angle, diving steeply with full flaps, he turned to me and said: 'This is what Antarctic flying is all about!' I knew what he meant. Professional pilots, condemned to a lifetime of sedate manoeuvring with loads of anxious passengers, are sometimes seen at small airports throwing light aircraft around the sky. They had taken up flying for the love of it, only to discover that earning a living by it brought hour upon hour of autopiloted tedium.

Two German aircraft landed to refuel at Rothera on their way south to their coastal base at 8°W on the far side of the Weddell Sea. One was a Dornier 128 and the other a Dornier 228, both with combined ski-wheel landing gear like our Twin Otters.[3]

Garry flew Dick Laws, Eric Salmon and me to Marsh on 13 January. I had learned a lot through serving as unofficial co-pilot to Garry for 80 hours' flying in six weeks.

Any lingering embarrassment about the Chilean invitations disappeared when the three of us were put aboard the FACH C-130, each hugging a full case of *pisco*, presented as a mark of their esteem. It took little more than two hours to reach Punta Arenas, quite a contrast with the slow ship crossings of Drake Passage that we had all endured in the past.

Although we had onward airline tickets, we were hustled aboard a FACH Boeing 707 and flown to Santiago, then led to the VIP lounge. The following morning Dick and I were ushered into the presence of the Chief of the Air Staff, General Carlos Desgroux, with General Lopetegui in attendance.

When these days of royal treatment by FACH finally came to an end, I left Santiago, wondering whether there was anything more that we could do to help our Chilean friends. Their gratitude was heart-warming.

18.   BARE BLUE ICE (1986–87)

With the world's leading climbers

Early in 1986 I was allowed one more fling in the Antarctic before reaching compulsory retirement age. I was to visit field parties to see if all was well and, if not, to consider what could be done about it. It was frustrating to be watching rather than doing; but, as usual, it was stimulating to be among young people at the start of their careers.

Ten months later, on my sixtieth birthday, I was cast to the winds – or a rocking chair – and given a pension. It had been a privilege to serve BAS as Head of Earth Sciences and I felt sad at leaving such a stimulating environment. In 12 years I had been able to transform the division into a fully professional group with a high reputation for the quality of its research. Starting with a handful of glaciologists on short-term contracts, I left behind ten teams of specialists each headed by a professional scientist on the permanent staff.

Only days after my farewell party at BAS headquarters, Giles Kershaw got in touch. I had not worked with him since our marathon radio-echo sounding season in early 1975 when we flew 214 hours in 6 weeks, much of it at low level. That record has never been matched since – at any altitude – by any pilot in Antarctica. From the start of Giles' time with BAS, we saw that he had a special talent for 'bush' flying – if one can apply the term to the Antarctic a thousand miles from the nearest bush.

In the course of the next 15 years he became the most accomplished pilot in Antarctic aviation history. On leaving BAS in 1979, he became free to fly non-government aircraft to a continent where only government aircraft were made welcome. The official line rested on the assumption that only governments had the means to operate safely. In the event of an accident, search-and-rescue operations could disrupt government activities. There were good reasons to be wary of accidents. Search-and-rescue activities following the crash of an Air New Zealand DC-10 on Mount Erebus in 1979 severely disrupted the US Antarctic Program for that season.[1]

Our paths had crossed again at Rothera on 1 January 1980, when Giles was employed as a Twin Otter pilot with Ranulph Fiennes' Trans-Globe expedition.[2] That was the private expedition that circumnavigated the globe via both poles. Although operating with only one aircraft, Giles arranged his own back-up options, and once turned the tables by saving the lives of three members of the South African government expedition who had strayed from their station without suitable equipment or experience.

When not flying in the Antarctic, Giles flew in Greenland and made many landings on pack ice in the Arctic Ocean using the Tri-Turbo, the hybrid three-engined DC-3. I had seen it in Punta Arenas in November 1983, southbound with a bunch of mountaineers, and again at Rothera three weeks later, northbound.

When the Tri-Turbo was no longer available, he continued to take climbers into the Ellsworth Mountains with a Twin Otter on charter from Kenn Borek Air of Calgary. The aircraft flew under the banner of Adventure Network International (ANI), a Canadian company set up for the purpose. The flying arm of ANI became the world's first commercial Antarctic airline.

Giles and I had known since 1975 about bare ice, or 'blue ice' areas in the Ellsworth Mountains. Although irrelevant to our activities at the time, it occurred to each of us independently that the ice might be smooth enough to take conventional wheeled aircraft, avoiding the need for skis. My view was that the feasibility of an ice runway could only be judged by careful examination on the ground (in this case, the ice). It was only after this interval of 12 years that Giles felt in a position to invite me to lead a joint Canadian–Chilean expedition to reconnoitre the area.

Now, eight days after retiring from BAS, I found myself on another Antarctic expedition. Some of my colleagues felt that this was an ill-judged switch from the public to the private sector. However, my view was that if blue icefields proved to be suitable for wheeled aircraft, government operators had as much to gain from them as anyone else. I knew that throughout Antarctic history, most logistic innovations have been the work of private expeditions. Governments generally prefer well-trodden paths even when new methods could be more cost-effective. Giles had hit upon an idea that might benefit everyone.

The expedition was being paid for by mountaineers heading for Vinson Massif. I met two of Adventure Network's employees in the airport at Miami. Mike Maxwell was a Canadian geophysicist hoping to combine studies of geomagnetism with whatever else ANI asked him to do. Hector MacKenzie was a native Scot, educated at Gordonstoun, who had emigrated to Canada and settled in the Yukon. Hector would lead a climbing group.

On arrival at Santiago, we were met by General Lopetegui, now retired from FACH but still pursuing his policy of assisting access to Antarctica to the ultimate benefit – in economic and political terms – of his country. Continuing to Punta Arenas, we met up with Giles Kershaw and were taken to a house that ANI had rented. Climbers were everywhere, some sleeping on beds, some on the floor, and others packing rucksacks. The President of ANI, Martyn Williams, was already in the Antarctic, but his partner, Maureen Garrity, was serving as manager of the house. We dined out with a group of Americans on their way north after successfully climbing Vinson.

The following day I was introduced to the rest of Giles' next Twin Otter load. I recognized Reinhold Messner, an Italian from South Tyrol and perhaps the most celebrated mountaineer of his day. With him were Wolfgang Thomaseth and Oswald Oelz. Each of them had paid £9,400 for the trip, in addition to the air fare to Punta Arenas. Hector MacKenzie was to guide two others, Reinhold Ullrich and Robert Cedergreen, an MD hailing from California. They had paid £12,000 each, to include the cost of their guide.

We were to share the Twin Otter cabin with their skis, ice axes, ropes, sleeping bags, tents and food. Most of the space in the cabin was already taken up with two large cylindrical fuel tanks each holding 950 litres. Without the cabin fuel it would be impossible to fly over the huge distances we had to cover. With it, I wondered whether the passengers were to be stacked one on top of the other. If that was the case, I would try to be on top of the heap.

It was 1 December before Giles considered the weather was flyable. Somehow, everything was crammed in and we were airborne at 0921. The acting co-pilot was Ron Kerr, a Canadian aircraft engineer. We were heading for Marsh, the Chilean station on King George Island, where Lopetegui had arranged for us to refuel. The airline distance from Punta Arenas to the foot of Vinson Massif is 2,850 kilometres, so if the Chileans would not help, our mission would have been impossible. Flying between cloud layers, we saw little of Tierra del Fuego and even less of Drake Passage until, through openings below, we spotted some loose pack ice. Scrunched in the cabin, sitting or lying on whatever we could find, the four-hour flight passed slowly. Toilet facilities comprised a communal open bucket.

At Marsh we were given a friendly reception and entertained to lunch with wine. The Chileans helped Giles with refuelling. There were three FACH Twin Otters on the ground, two of them sheltering in a sizeable hangar. It was sunny and warm with the air temperature 0°C. After consulting the local meteorologist, we took off for BAS's abandoned Adelaide station – the Chileans had renamed it Teniente Luis Carvajal.

This was a relatively short leg of 780 kilometres. However, we ran into problems, beginning with ice accumulating on the airframe. Giles climbed to 4,300 metres in an attempt to get above the icing conditions, but to no avail. The only alternative was to go down. When we finally broke cloud we were 30 metres above the sea. I recognized the ice-cliff coast of Adelaide Island but had no idea where along its 200-kilometre length we were. Giles flew over the island and recognized a couple of nunataks. None of my colleagues looked at ease as he again plunged into cloud, climbing, only to emerge just above the sea because the ice we were carrying refused to go away. I was probably the only person on board who knew, from experience, that Giles could handle the situation.

I remembered reading in the Twin Otter flight manual: 'This aircraft must not be flown in icing conditions.' However, at this stage we had no choice. Following the coast at 20–30 metres over the ocean, we never did shake all the ice off and landed with some still clinging. We were met by Captain Moreno Mora, who, I remembered, was with Lopetegui and me when we came here from Rothera in a FACH helicopter three years before.

Moreno invited us down to the station for rest, meals, beds or anything else we needed. The station itself looked in good shape and quite well maintained. A welcome addition to the earlier BAS facilities was a flushing toilet. Moreno was smiling and cheerful, the only officer among six unsmiling airmen. I surmised that he might be a volunteer whereas they were conscripts. On the other hand, Chilean airmen are generally rather deferential in the presence of officers.

Unfortunately for old folks like me, Giles was not the resting kind. The aircraft was costing ANI £22,500 per month in addition to an hourly rate for flying, so rest was a luxury he could not afford. After soup and spaghetti, he led the way up the hill to the airstrip. There was a rising wind and drifting snow, but in spite of it we took off. This last sector of the flight had to cover 1,300 kilometres. Again we flew between cloud layers until the great Sentinel Range loomed ahead. Here we had to climb to 4,500 metres to get over the range to the place where ANI had its camp. Tiny tents to the left of us showed that there was a climbing party on the mountain, but at this time of night they would be fast asleep.

Dropping steeply to the area of the main camp at 2,300 metres, we could see nothing, the camp was in fog. I wondered how Giles would deal with this one. Undeterred, he landed on a mountain col above the fog and slithered downhill, using reverse thrust to maintain control. I wondered how many other pilots had ever tried downhill skiing with a million-pound aircraft. When we taxied into camp, the visibility was less than 100 metres. Giles apologized for the manner of our arrival, adding: 'I promise that we were in control *most* of the time!'

I had a headache as a result of 13 hours in the air, much of it above 3,000 metres. I found an empty tent and fell into a sleeping bag 24 hours after getting out of bed in Punta Arenas. Mike Maxwell stayed up for some time to deploy a pair of fluxgate and induction magnetometers for a joint project of the universities of British Columbia and Tokyo. This was ANI's first attempt to attract genuine research projects. Mountaineers did not need encouragement.

Breakfast was prepared in a 2 by 4-metre 'Weatherhaven' tent by whoever first became hungry. Giles was usually hungry and also, I knew, a very good cook, so he was to be seen stirring the pot more often than anyone else.

The route from base camp to the summit of Vinson at 4,897 metres was well trodden, with three intermediate camps on the way. We were still in fog, but that did not deter the climbers. Ullrich and Cedergreen were the first to

The Sentinel Range with Vinson Massif (4,897 m), the highest mountain in Antarctica

leave, on foot, with a depot load for Camp 1. MacKenzie followed them on skis. Then Messner and Oelz left on skis with quite lightly laden rucksacks. Thomaseth filmed their departure and then followed. Some hours later the first three returned and, after a meal, started up again. Ullrich was on skis with a large rucksack, the other two on skis towing small plastic sledges. The contrast between the light loads of Messner's party and the heavy loads of the others spoke volumes about Messner's estimation of how long it would take him to polish off the mountain. By his standards, I suppose, it was a pygmy of a mountain. However, I recollected that, three years earlier, Chris Bonington told me that he himself had underestimated the climb.

Martyn Williams was a middle-aged renegade from Wales who had set up a business in the Yukon to lead climbs and wilderness treks. That evening we spoke by radio with his party on the mountain. They had successfully summited and were on the way down. With him was Alejo Contreras, a Chilean climber who was serving as liaison officer between ANI and FACH.

Hector MacKenzie reported that all was well at Camp 1. Messner's party, however, had no radio. With long experience as a solo climber, he despised any weight that could be dispensed with.

The next morning we were still in cloud but Giles taxied down to the local fuel depot. This was a supply of avtur that had been airdropped from a FACH C-130 some weeks before. In previous years, ANI had paid the full cost of Chilean airdrops, together with parachutes, cargo pallets and the fuel itself. This meant that the cost of fuel here was £2,060 per drum – £10.30 per litre – about the same as whisky at home.

On this occasion, however, the whole cost had been borne by FACH in exchange for whatever I might find out about the possibility of landing on blue icefields. This placed rather an onerous burden on me because, at this stage, it was no more than an act of faith.

Late in the evening, Martyn's eight-man party loomed out of the fog in ones or twos over a two-hour period, every one of them looking exhausted. Martyn himself was carrying a heavy pack and also dragging a sledge – he must have been fitter than his flock. Giles prepared dinner and all of us sat talking until midnight.

The sun shining on my tent the next morning gave promise that the time had come for me to be useful. The sky was clear and I enjoyed my first good view of the mountain since we had arrived. Five of us took off on the long-heralded search for blue icefields: Giles was pilot and I was co-pilot. It was more than a decade since I had sat beside him in this capacity, but the years fell away and we at once dropped into the old routine. 'Steer one three zero,' I said, 'estimate abeam Welcome Nunatak zero nine three zero.' With us were Ron Kerr, Alejo Contreras and Mike Maxwell. If we found any blue ice worth surveying, Mike and I would camp on it for a few days. Mike had prepared the camping gear. On both sides we had mountain vistas of sur-passing grandeur. Once again I felt what a privilege it was to be here and what fun it was to be beside my old friend.

I had prepared carefully, conscious that for every single minute of flight we were burning £50 worth of fuel. Having studied Landsat images to iden-tify every possible blue-ice area, I directed Giles on the most economical route to pass near all of them. After a little over an hour we were over an icefield at Wilson Nunataks. I measured it by timing a low pass from end to end – it took 53 seconds at 120 knots. This was good enough, so I said, 'Let's land!' It took some minutes for the hydraulic system to raise the skis so that the wheels could project beneath. I was apprehensive on the final approach. On touchdown we felt a high-frequency vibration as the wheels

Visiting a group of geologists in Palmer Land

Eric Salmon beside the milepost at Bellingshausen station

Martyn Williams

First landing at Patriot Hills. The skis are not touching the surface

Giles Kershaw cooking for an exhausted climbing team returning from the summit of Vinson Massif

Mike Maxwell preparing beef steaks for dinner

Reinhold Messner sets out for the summit followed by Oswald Oelz

A refuelling stop on the way home. Left to right, standing: Ron Kerr, Reinhold Ullrich, Oswald Oelz, Reinhold Messner, Charles Swithinbank, Giles Kershaw, Hector MacKenzie. Sitting: Bob Cedergreen, Mike Maxwell, Wolfgang Thomaseth

ran over a wind-scalloped surface that we had seen from the air. The bumps were only about 5 centimetres high, thus easily absorbed by the low-pressure tyres. Giles said that they actually helped with wheel braking. We got out and walked around. Apart from the scallops, the surface was incredibly smooth. I concluded that this spot was usable for transport aircraft but that we might find something better elsewhere.

Now we flew to the southernmost part of the Heritage Range. There were long stretches of blue ice at the foot of Independence Hills but the place which really took our fancy was a 2 by 8-kilometre area of bare ice off a ridge called Patriot Hills. This was the one that Giles and I had glimpsed from a distance 12 years before. We had both noticed how smooth it seemed.

This time there was a strong and turbulent wind blowing over the ridge, and I could see that Giles was becoming itchy about his fuel state or perhaps the rate at which we were sapping the Chilean economy. There was snow on the surface but we could see that it was not more than 15 centimetres deep. As we landed, the wind and the snow between them acted as if we had snagged the arrester wire of an aircraft carrier. Mike and I unloaded the camp gear and waved goodbye. Giles tried to turn downwind to return to the spot where we had touched down, but the wind was having none of it. Each time he tried to turn, his big vertical stabilizer acted like a weathercock and skidded the aircraft straight again. He had to take off aiming straight at the mountain wall, but managed to clear it with a steep climbing turn.

A bleaker place would be hard to imagine. The wind was cutting through our clothing. A few small bundles at our feet were all that stood between us and encroaching hypothermia. We pitched our tiny mountain dome tent but tent pegs were useless because the surface was like concrete. Mike had brought ice screws, so we laboriously screwed them into the ice, one by one, and attached the guys. All my life I had enjoyed the luxury of pyramid tents that felt safe in almost any wind. Now I was to live in a survival tent made for backpacking mountaineers who did not seek comfort.

Mike built a protective wall of snowblocks on the upwind side. Climbers do it, but elsewhere in the Antarctic it would be a recipe for burial under a snow-drift. Even in the lee of the wall, the tent was flapping wildly and I thought of our predicament if it should tear away. We strung an alpine rope criss-cross over the dome and secured the ends with ice screws. Now I felt safe.

We had brought a small chainsaw to cut ice blocks for drinking water and to ensure that there were no voids beneath the surface that might collapse under the weight of a 100-tonne aircraft. It worked fine but the chain oil imparted an unusual flavour to our drinks.

Although more than 99 per cent of the surface of the Antarctic ice sheet consists of snow, the existence of 'blue icefields' was well known.[3] They occur in places where wind and sublimation (or evaporation) strip off the snow which covered the ice before it flowed into a localized area of mountain turbulence. However, the process can be very slow indeed. Some aluminium survey markers that I myself drilled into the surface of a blue icefield in 1951 were found still standing 34 years later. Changes in the exposed length had averaged less than 2 centimetres per year.[4]

My plan was to use an engineer's level together with a 4-metre graduated staff to measure the slope of the ice and the length of any potential runway. We tried to begin as soon as the tent was secure but the wind threatened to topple the tripod, so we gave up and retreated inside.

Mike had brought a radio and now spoke with Ron Kerr. He and Giles were already on their way to Chile with all the climbers who came down from the mountain the day before. If they continued straight through to Punta Arenas, they would be flying for not less than 13 hours. Alejo had been left alone at the base camp, and we spoke with him too.

As the sun fell behind Patriot Hills, we began to hear rifle shots. Before surrendering to any attackers, I remembered that I had heard it all before. At nightfall on blue icefields, intense outgoing radiation cools the surface, contracting the ice until it fractures. Sometimes a fracture propagated through the ice on which we were lying and I had dreams of being swallowed up in a newly opened crevasse. But the barrage died down and after a supper that Mike prepared on a camp stove, we slept well.

In the morning, in spite of a 20-knot wind, we spent the whole day surveying what looked like a potential runway long enough for any aircraft yet built. But it was across the prevailing wind, so would not be popular with pilots. An into-wind runway would have to head straight for the mountain, and that would be even less popular with pilots.

The procedure was for Mike to move ahead 50–100 metres with the staff, and to hold it straight up and down while I peered through the telescope of the level and read the graduations. Then I would fold the tripod and walk past him to take a back-reading on the staff. This is the conventional way to compensate for refraction, which under these conditions can be extreme. Mike would then move past me to a new position, all the time keeping aligned in the direction of our 'runway'. By the time we were frozen and I was sunburned, we had covered 2,300 metres. It was only when we got inside the tent that I realised just how badly sunburned I was. My lips were swollen, my nose was swollen, and my eyes were half closed from swollen eyebrows. From that day on I had reason to believe in the ozone hole and its effect on the amount of ultraviolet radiation reaching the ground. We were not far from the geographical centre of the hole and only a few weeks had elapsed since its maximum extent. Mike had a tougher skin than mine and was just tanned.

In view of the cost of the operation, there was no way in which I could plead sickness, so we must carry on. Raiding the first aid kit, I found a triangular bandage and next morning made a mask fit for any bandit. Above it

176

were my dark glasses and above those my cap pulled down. Nothing was left exposed. Faced with a miserable 20–30-knot wind, I had to hold the tripod with one hand before propping it against my back to write down the numbers. Finishing the last bit of a 3,500-metre runway, we retreated to the tent exhausted. Mike heard that all the climbers on our own flight south had returned to the base camp after successful climbs. At nightfall the wind blew up to a real blizzard, but by that time we were snug in our sleeping bags.

On 7 December, Pearl Harbor Day, it was overcast, snowed all day, and blew at 10–20 knots with an air temperature of only –5°C. Much of our icefield disappeared under a blanket of new snow. In visibility down to 100 metres, we ventured out for a couple of hours to do a runway micro-relief study. This involved determining relative levels every 2 metres to see if there were any short-wave bumps that might jolt a speeding aircraft.

I concluded from our work that there was a potential 3,000-metre runway which appeared adequate for transport aircraft. The big problem was turbulence caused by migrating vortices coming over the ridge on the upwind side. Almost all the snow falling on the ice had been swept away by the time we left. Later we discovered where it had gone. It had been dumped 1 kilometre downwind – in areas where the turbulence had moderated. The gradient of the runway was one per cent, or about half a degree, which was well within published runway criteria for transport aircraft.

Two days later, Giles came to collect us. As co-pilot he brought Captain Alejandro Frías, who had arrived at Vinson base camp in a FACH Twin Otter. The Chileans were taking a very active interest in our doings. Frías was invited to handle the take-off facing into the mountain wall and he did not seem too happy about it – but there was no other way out.

The base camp was socked in, so we had to repeat the performance of landing higher up and then skiing the aircraft down to the camp. When the camp loomed out of the fog, we saw two FACH Twin Otters neatly lined up. Evidently they had come in pursuit of Lopetegui's policy of learning from us about high latitude operations. I gathered that one of the pilots, low on fuel, had been alarmed to arrive in the vicinity of the camp without seeing it. Giles had suggested over the radio that he should land anywhere and wait for better weather. The pilot saw that as an emergency situation and therefore to be avoided, whereas Giles saw it as the best way to avoid an emergency. Giles was right.

The Chilean aircraft landed above the camp, as we had, and then skied down. Giles' tracks were fairly straight but theirs looked more like uncontrolled slalom.

The visitors were camping in small dome tents. Unused to temperatures of –20°C, they had brought with them a small generator to run electric heaters in their tents, then kept it running all night. This did much to explain why they were now even lower on fuel. However, miserable as they appeared, they were as friendly as ever and grateful for anything that we did to make their stay more tolerable.

Bob Cedergreen, the MD from California, was fascinated by my sunburn-swollen face and invited me to sit for a portrait, perhaps soon to grace some learned journal of pathology. Luckily for me, the swelling rapidly subsided.

Giles was anxious to get going. On 10 December the 'runway visual range' was 1,000 metres in *any* direction – as was the 'runway'. Considering the altitude of the camp, the aircraft was heavy, so take-off was something of an act of faith. After skiing downhill at full throttle for what seemed like miles, all of us in the cabin were willing ourselves into levitation. Finally, the machine staggered into the air. There were ten of us on board with all our equipment plus the two giant fuel tanks.

Five hours later we landed at Rothera. Knowing BAS's policy with respect to private expeditions, goodwill is all we sought. I offered to carry their mail to Punta Arenas, so there was plenty of goodwill to be found. The adventurous members of the station found it a special privilege to meet Reinhold Messner and his colleagues. Giles and I, both ex-Fids, were of course among friends. From Rothera it was a 15-minute flight to Carvajal, where we refuelled from eight rusty fuel drums.

At Marsh we were asked to add a sick Chilean to the passenger list. Considering what the Chileans had done for us, Giles could hardly refuse, so the flight to Punta Arenas was more crowded than ever. It was one of many examples of the unwritten laws of human kindness in Antarctica.

The dawn was breaking over a windy Patagonia when we finally landed, 23 hours after leaving Vinson camp.

Five days later in Santiago, General Lopetegui ushered me and Martyn Williams into the spacious offices of General Fernando Matthei, Commander in Chief of Fuerza Aérea de Chile. Matthei was a member of the ruling military junta of the country under General Augusto Pinochet. Pinochet was getting a bad press outside Chile – notably in the USA – but I found Matthei to be a gentle, charming and highly intelligent man. He spoke fluent English and was interested in the possibility of taking his own wheeled aircraft to the Patriot Hills icefield.

By now I was convinced that our icefield could take transport aircraft of any size. However, it was understandable that neither Chile nor any other government wanted to commit their aircraft to landing on ice until the concept was proven. It was in Matthei's interest that ANI – and not FACH – should bear the risk of a proving flight.

On reaching home I wrote a formal report to ANI and FACH on the merits and drawbacks of an ice runway at Patriot Hills.[5] The FACH copy landed on General Matthei's desk. Now it was up to ANI and FACH to chart their own course.

19.   PIONEERING (1987–88)

We teach an airliner to pirouette

My 16-page report on the 1986–87 season's work included a recommendation that 'Final site selection [for an ice runway] should be based on a large-scale map of the icefield with 0.5 m contour interval.' General Matthei had accepted this and delegated the task to Servicio Aéreo Fotogrammetrico. I had visited them on 16 December 1986 and arranged for the mapping to be done.

However, my diary entry for that day revealed some doubt:

> What problems they will find along the way – they always do – time alone will tell.

I was more than a little surprised to hear from ANI shortly after they received my report in March 1987 that: 'Kenn Borek Air is moving along with the acquisition of a DC-4 for next season'.

The DC-4 is an ancient four-engined piston-driven airliner with a strong landing gear and a slow landing speed – but no skis. Whatever aircraft they chose had to be able to fly an unrefuelled distance of 6,400 kilometres, with reserves, while carrying a commercial payload of four tonnes. The round trip distance between Punta Arenas and Patriot Hills is equivalent to flying non-stop from London to New York. With a wheeled aircraft, the only possible diversions en route would be to Marsh or Marambio. A DC-4 was the only affordable type that could satisfy these requirements.

It was not long before ANI invited me to go with the first flight and to stay on the ground for a few days. I thought long and hard before accepting because, if I refused, they would go anyway. If I went, there was a chance that I could stop them doing something unwise, so I consented.

I prepared memoranda on how to behave on an icefield, particularly on the question of waste disposal. Over most of Antarctica, snow accumulation conceals forever anything left on the surface. Not so on an icefield. I wrote:

> Waste left there will remain exposed for eternity. It will be all too easy to foul the nest with the effluent from human occupation
> . . . No Antarctic station in history has ever been situated in an internally-draining bare ice area. We are breaking new ground. The eyes of the world will be upon us. If we mess it up, the wrath of the world will be upon us.

ANI had settled on an air armada consisting of one DC-4 and two ski-wheel Twin Otters. The only way that they could ever cover the enormous costs of chartering three aircraft was to take in money from clients in advance. This involved promising to carry 60 people to Patriot Hills, then flying about half of them to the foot of Vinson Massif and the other half on a short visit to the South Pole. The South Pole passengers had to pay between £15,000 and £21, 000 each, the higher figure being for the first flight. To a few people, it meant that much to be the first tourists ever to stand at the South Pole.

I reached Santiago on 5 November 1987 – General Lopetegui was at the airport to meet me. He said that the US Government was trying to prevent the DC-4 from leaving Chile for the Antarctic, on the grounds that such operations were hazardous and would probably involve the US in search and rescue. The State Department in Washington, possibly at the instigation of the National Science Foundation, was acting through the US Ambassador in Santiago. The Ambassador asked General Matthei to refuse permission for the DC-4 to proceed to the Antarctic. Under Chilean law the General had power to do this.

However, ever since our meeting with him last season, Matthei had been on our side. I never found out what words he used to give the US Ambassador the brush-off.

The American dislike of mountaineering expeditions seemed to be focused on *foreign* expeditions. In 1966–67 there had been an American Antarctic Mountaineering Expedition that was not only allowed to climb in the Ellsworth Mountains but also flown there in an LC-130 at taxpayers' expense. The ten members were selected by the American Alpine Club to ensure that the first ascent of the continent's highest peaks could be credited to Americans. Members of the expedition attained the summits of five peaks including Vinson Massif.[1]

I arrived at Punta Arenas to find both ANI Twin Otters at the airport. The DC-4 was delayed in Lima, Peru, needing an engine change. Several of last year's party were in town: Martyn Williams, Maureen Garrity, Hector MacKenzie, Mike Maxwell and Alejo Contreras.

Before I left home, ANI had asked me to prepare a few lectures on various Antarctic subjects for the benefit of their staff and also the first group of Vinson climbers who were already in Punta Arenas. Over the next few days I gave illustrated lectures on:

The Ellsworth and Thiel Mountains
Antarctic aviation
The Antarctic Treaty and its relevance to private expeditions
Antarctic scientific stations
Antarctic ships

The first of the two Twin Otters left for the Antarctic on 12 November, staging through Marsh and a fuel depot on Jones Sound, not far from Rothera. It carried Hector and the first group of Vinson climbers. The DC-4 eventually arrived on 15 November, the crew tired but triumphant. The Captain was Jim Smith, an old Arctic hand with thousands of hours of DC-4 time behind him. With him was a second captain, Bob Craig; their young co-pilot, Brydon Knibbs; and two engineers. They had brought five tonnes of cargo. We spent the rest of the day sorting it with the able assistance of a FACH officer as well as cargo handlers from the state airline LanChile. The help freely given from all sides was quite embarrassing.

The second Twin Otter left on 15 November with Andy Williams, Mike Maxwell and Martyn Williams. Like Martyn, Andy was a renegade Welshman who lived in Whitehorse and served as pilot for the Arctic Institute's Kluane Lake camp. He had earlier wintered with BAS as radio operator at Halley Bay. Now he was to be Camp and Construction Manager for Patriot Hills.

The first work for the DC-4 was to carry a load of avgas to Marsh in case fuel was ever needed there for an intermediate or emergency landing. I was asked to go along. The only other passenger was a Chilean helicopter engineer. The last time I had flown in a DC-4 was 30 years before in Canada and I had forgotten what an amazing workhorse it is. This particular aircraft, N4218S, was 44 years old and had flown 37,000 hours. On one side was painted ANTARCTIC AIRWAYS and on the other side ADVENTURE NETWORK INTERNATIONAL. A forklift hoisted 50 drums into the cabin, a total weight of more than eight tonnes. Besides the cargo, we took off with fuel for ten hours' flying.

We were not airborne until six minutes after the airport had officially closed for a military exercise. 'Playing war!' as a sarcastic airline pilot had put it. Cruising at 160 knots, we saw the South Shetland Islands ahead after three and a half hours. A small crowd watched our landing at Marsh and we presumed it was just a friendly welcome. It turned out that we were the first DC-4 ever to land there. Some Chileans had predicted that we would fail to stop before the end of the short runway or perhaps sink into the mud. Before we had time to ask, a forklift was waiting to take our 50 drums to a storage dump. ANI's second Twin Otter had landed shortly before us and the crew were busy digging out avtur supplies before continuing south. The CO of the base, Commander Juan Bastías, lit up with delight when he saw that I had brought a bag of mail. He had been here for a year and had another year to do. However, his family was with him, so he had no hankering to go home. The settlement had expanded since my last visit and now included 13 families of officers in separate houses. Twenty-three children, aged between 1 and 16, had their own small schoolhouse with a resident teacher.

We were asked to take an injured helicopter pilot back to Punta Arenas and readily agreed. There was a waving crowd to watch him leave. They must have been as surprised as I was at the very short take-off run of the now empty DC-4.

At Punta Arenas airport there was a brand new British-built Islander aircraft on a delivery flight to the Falkland Islands. The Argentines had refused to allow it to fly over Tierra del Fuego or indeed anywhere in their airspace. The passage of five years since the Falklands war had done nothing to diminish their antagonism to anything British. What a contrast were the Chileans! The pilot was offered a FACH jet fighter escort until he was handed over to RAF aircraft operating out of the Falklands.

Giles Kershaw arrived in town, having been given a few days off from his job as First Officer of a Boeing 747 belonging to Cathay Pacific Airways. We were about to take a momentous step in the evolution of ANI and he was determined to be a part of it.

The great day came on 21 November. The load in the DC-4 was frightening to behold. We had fuel on board to allow for 24 hours in the air. Not only were the normal tanks full to capacity but a pillow tank in the cabin held 4,550 litres. A separate cylindrical tank held another 950 litres. Then there were 17 drums of avtur for the Twin Otters, each weighing 180 kilos, and a few tonnes of camp gear. There were three other passengers – more properly described as supernumerary aircrew because this, being a fuel flight, was only supposed to carry aircrew. My companions were Giles Kershaw, Jill Pangman and Lis Densmore. Jill was a biologist masquerading as dietician – a euphemism for camp cook – because she desperately wanted to go to the Antarctic and ANI had only this vacancy. Lis was to serve as camp doctor. She too was so keen to go south that she had left her two children with relatives in Whitehorse.

There were three seats at the front of the cabin. Eyeing the load, however, I decided that, in case of any mishap, I would prefer to be behind the load during take-off. Not wishing to admit what I was thinking, I quietly sneaked aft and, facing aft, cushioned my back against the load. I was discovered when Captain Jim Smith came aft for a final check before take off. I blurted out: 'I would prefer to stay here for the moment, if you don't mind.' He knew at once what I was driving at, then smiled and left me.

I think all of us held our breath as we trundled down the runway at Presidente Carlos Ibañez airport. After what seemed like an age, I looked out of a window to see the perimeter fence flash by not far beneath us. I wondered whether any DC-4 had carried this much since the Berlin airlift.

It took half an hour of climbing to clear the peaks and glaciers on the south side of the Strait of Magellan. Turbulence caused avgas to belch from the overflow of the cabin tank and it took no feat of intellect to see what could happen if the fuel ran down into some electrical circuit and caused a spark. When I told him, Jim acted extremely fast to draw down the fuel in the tank. Another problem exercising the crew was a tendency for the mountain turbulence to induce stalling. However, all these difficulties disappeared as the engines burned down the weight of fuel.

As a precaution, Jim diverted eastwards to make landfall over the Antarctic Peninsula. It added distance to the flight but gave comfort that we knew where we were. Giles took a turn in the left-hand seat as we approached Alexander Island. A low ceiling was keeping us well below the mountain peaks. After discussing our options with me, he chose to fly up Hampton Glacier. Having flown it more than once, both of us knew this was a wise choice. However, our colleagues saw half-clouded mountains flashing by on either side and wondered how the ascent would end.

Spirits soared when, 11 hours out of Punta Arenas, we saw the whole sweep of the Ellsworth Mountains spread out on our right side. I did not need a map and just pointed the way. Jim took his place in the left-hand seat, Bob Craig in the right-hand seat. I had expected them to want Giles to be there. But no, I myself was summoned to sit in the jump seat between the pilots. The implication was clear: if the landing ended in disaster, Jim wanted me to share the blame.

Letting down as we approached Patriot Hills, we were surprised to find that the icefield was not an empty spot in the wilderness. Both ANI Twin Otters were tied down near two FACH Twin Otters. The Chilean Twin Otters had come all the way from Marsh, 2,100 kilometres away, presumably at the behest of General Matthei. Without an airdropped fuel depot at Welcome Nunatak, they would never have made it. All four aircrews and ANI's camp staff were lined up to witness the landing.

The ground party reported high and gusting winds. Heavy snow was drifting over the ridge and whirlwinds were spinning across the icefield. Several times Jim needed to use full aileron to stay upright. I explained that I would not tell him where to land but I might well tell him where *not* to land. He could not use the long dimension of the icefield because it was beyond the crosswind limits of the aircraft. We would land straight towards the ridge, an approach that is normally against any pilot's instinct. I estimated that there was 1,000 metres available between the end of the bare ice and a moraine ridge up against the mountain. In view of the conditions, I was quite expecting Jim to say, 'No, we're going home.' However, with the wind blowing over the ridge at 35 knots, he felt that he could stop in time.

With icy calm he and Bob rattled off the landing checks: 'Gear down, flaps thirty, three green . . . ' I had suggested that in order to slow us down, we should hit the ground before the snow-cover gave way to bare ice. Touchdown was exactly in the right place. A series of crashing jolts as we ploughed through sastrugi helped to bleed off some of the speed. After this we were on bare ice with the expected high-frequency vibration from the scalloped surface. Jim hesitated to use wheel brakes because of the possibility of skidding. My diary continues:

Patriot Hills (centre foreground). Aircraft land on the undulating stretch of bare ice beside it. Vinson Massif, 225 kilometres from the camera, looms on the horizon (top right)

Now running downhill straight at the Chilean Twotters, I had visions of ploughing into them and ending up a twisted heap in the moraine.

But we slowed in good time. The wind was so strong that when Jim tried to turn downwind, the machine simply weathercocked, so we shut down. We were on the ice, intact, and safely parked. Congratulations flowed from all sides. Ten thousand old rivets still rattled as the aircraft rocked from side to side in the wind. The Chileans were all clad in eiderdown suits and we could see why.

There was no time to lose because if the engines cooled too much they might refuse to start. The air temperature was –25°C but the wind-chill factor made that equivalent to –55°C. Everyone joined in the unloading. Meanwhile the engineers had the most miserable task of all. One of them had to walk out on the wings with a hose so that fuel could be pumped from the cabin into wing tanks. The wind sprayed avgas all over his clothing and he needed help to prevent the hose being blown out of his grasp. His wind-chill factor must have gone right off the scale.

One of the Chileans, Alejandro Frías, had come here the year before when Giles collected me and Mike Maxwell at the end of our survey. Another was Major Patricio Toro, the C-130 pilot who had flown me, Dick Laws and Eric Salmon from Marsh to Punta Arenas in January 1984. Toro's inclusion in the FACH Twin Otter party told its own story. He had been sceptical about the ANI operation but was now convinced. He jumped aboard the DC-4 to congratulate Jim Smith and promptly asked to hitch-hike back to Punta Arenas. The alternative of staging slowly northward with the FACH Twin Otters did not appeal to him. He declared his intention to return to Santiago and fly his C-130 here. In the event, it took several years to convince the authorities.

Now on the ice, I sheltered behind the fuel drums with the ANI party while the DC-4 taxied back to the threshold and turned towards us, seemingly awfully close. It must have been frightening for them to face straight into the mountainside but there was no choice. A mighty roar told us that Jim was opening up to maximum power before releasing the brakes. When he did, the machine bounded forward and leapt into the air after no more than 300 metres, an amazing performance for a plane that was carrying fuel for 12 hours' flying. They cleared the mountain face with a climbing turn.

The wind made it hell on the ground but it would have helped the pilots. I was reminded of the time, 42 years before, when I served in an aircraft carrier. One day a Seafire, the naval version of a Spitfire, took off in the space of 60 metres. Wind over the deck did the rest.

The ANI camp was on a broad moraine running along the foot of the ridge. The Chilean camp was 300 metres north of ours but they quickly headed home. Our advance party had erected several Weatherhaven tents and we were soon warming ourselves inside. Giles had chosen to return north with the DC-4. Jill and Lis combined to make excellent meals from a limited variety of ingredients.

The Twin Otters seemed rather precariously parked beside each other hard up against the moraine. I remembered the awful day in November 1981 when both BAS Twin Otters had been flipped onto their backs at Rothera by hurricane winds, so I suggested that one of ANI's should be picketed some way out on the icefield. Then if we did encounter freak winds at one spot, the other aircraft might escape destruction.

For days on end, winds continued at gale force. In calmer periods, some of the party climbed to the summit of Patriot Hills or went skiing on the slopes. Mike and Jill had brought skates and we watched them playing vigorous hockey matches on the surface of a frozen melt pond. Andy and Mike put out runway markers and the aircrews were kept busy with minor repairs. I did not envy the engineers having to wield spanner and screwdriver with bare hands.

At one end of the camp there was an aluminium mast with five colourful flags flying bravely in the breeze to celebrate the international make-up of the tiny camp crew: Chile (Alejo Contreras), Canada (several), Switzerland (Henry Perk, pilot), Yukon Territory (several) and UK (me). Nobody had planned a cosmopolitan team – it just came out that way. The climbers could have added another couple of flags.

After a few days Hector called on the radio from Vinson camp to say that the climbing party was down from the mountain and ready for pick-up. Henry Perk flew north with Alejo to bring them in. They reported one broken rib and lacerations from an uncontrolled slide – but otherwise all well. Lis, the doctor, was quite pleased to have something to do. However, she felt unqualified to deal with the most serious ailment – a case of injured pride. One of the climbers had failed to make the summit and was now blaming everyone in sight. He wanted to go home and was unable to understand why the DC-4 was not running a daily shuttle service to Punta Arenas. At one point he was caught trying to use the radio to call for an emergency evacuation flight. We had to put a guard on it.

Martyn had brought a slide projector and I gave several lectures to entertain the climbers. The 24-hour daylight did nothing to help the illustrations but a sheet hung on one end of the tent served as a passable screen.

On 30 November the DC-4 took off from Punta Arenas but the landing gear failed to retract. A heavy cargo load combined with the extra drag made for a rather hair-raising emergency landing. However, they tried again the following day. Airframe icing kept them low for part of the flight. Our weather deteriorated in the course of the flight with a low ceiling and white-out. I wanted to suggest that Jim should give up, but it was not my place to do so. All we could do was to tell him the situation over the radio.

When he did reach us, we heard him pass over a fogbound Patriot Hills heading towards the South Pole. However, seeing no nunataks ahead, he turned round and found us. The climbers, anxious to get going, walked over the icefield and lined up alongside the intended runway. I decided to stay well clear in case the aircraft skidded. The camp was in a hollow, so I could not see the actual touchdown. Rucksack on my back, I was walking up towards where I expected the plane would park. Only when I heard throttles being opened and closed in rapid succession did I realize that something was drastically amiss. The aircraft, obviously out of control, appeared over my horizon skidding in circles (in aviation terms a ground-loop). It had veered off the line of the runway and was charging downslope, rotating as it went, and coming straight towards me. I contemplated ducking between the legs of the landing gear, but the rotation would have made it difficult. So I turned and fled.

Some years later Martyn, who heard the story from the pilots, told me that he would retain for ever the image of an old man racing downhill to escape from a pirouetting airliner. It is etched as sharply in my own memory.

The DC-4 came to rest among scattered boulders beside the moraine. It had been a miraculous deliverance. Nobody was hurt and the legs of the aircraft still stood upright. A very worried aircrew came down from the cabin to inspect the landing gear. There were rips in the tyres and spongy patches where rubber had burned from the heat of friction when the wheels were locked. The alternating bursts of engine power I had heard were vain attempts to stop the rotation. All of us were consumed with admiration for an airliner that could survive an event like this.

My colleagues saw some poetic justice in the fact that I, the most cautious onlooker, had been put at greatest risk. We had a good laugh, which put this bizarre adventure in perspective. What really mattered was that our salvation, the DC-4, was still in flying condition – though perhaps only just.

After unloading the cargo, the climbing party mounted the ladder with Hector, Lis and me. We were about to undertake a ten-hour flight over hostile terrain in an aircraft which, by virtue of its tyres, would not be considered airworthy anywhere else in the world. I was again asked to sit in the jump seat between the pilots. In run-up of the engines at the threshold it was clear that No. 2 engine was not pulling as it should. We waited a while and luckily it recovered. Now I too had the experience of seeing the mountain alarmingly close ahead. As before, Jim held the brakes until all four engines were straining at full power, then suddenly let go. As we accelerated through 60 knots, the co-pilot, Bob Craig, called out: 'Number three engine fire warning!' I held my breath but Jim did not flinch. If he had aborted the take-off at that stage, we would have skidded straight into the mountain. So he chose to fly, banking sharply against the mountain. Three times I saw him hold full aileron to counter the vortex coming over the ridge.

The fire warning had been due to an oversensitive microswitch. When, some time later, I went aft, almost everyone was asleep. There was an air of calm resignation. Chris Briggs, the Chief Engineer, was a man after my own heart. He told me that on final approach to landing at Patriot Hills, he had taken the precaution of positioning himself in the safest place – behind the cargo. At some point in the ground-loop he realized that the aircraft was travelling backwards and he would be better off in front of the cargo. Relief at the outcome yielded many a laugh.

I was called up front again to say whether we were too far east over Palmer Land, which we were. There were high mountains in the vicinity which we could not see for cloud, so Jim climbed to 3,200 metres to be on the safe side. Finally we got a good fix over Spaatz Island and set course for Punta Arenas.

That was the end of my contribution to the season. Valuable lessons had been learned from the first two flights. Over the next few weeks, the DC-4 made 11 more flights to Patriot Hills without mishap. Under average wind conditions, the round-trip flying time for each excursion was 20 hours.

Twin Otter flights to the South Pole took place in January. Incoming DC-4 passengers were housed in the Patriot Hills camp while they waited for good weather to the south. The distance from Patriot Hills to the South Pole being 1,080 kilometres, an intermediate camp was established in the Thiel Mountains to provide a safe haven in the event of weather delays. Five round-trip flights safely carried a total of 32 passengers.[2]

The National Science Foundation was strongly opposed to tourist visits to the South Pole but had no legal means to stop them. Initially they refused one of the most basic norms of aviation – the provision of weather information. When the aircraft did arrive, the station's air traffic staff had been instructed to deny use of their runway. Having anticipated this difficulty, I had advised the Twin Otters to use it anyway. It may not have occurred to NSF that visiting aircraft, forced to land on rough sastrugi off the runway, would be more likely to precipitate the kind of emergency that they feared most.

FACH had formally agreed, on its own initiative, to provide search-and-rescue cover for ANI's operations. However, it was obviously in ANI's interest to do everything possible to avoid asking for help from anyone. I had insisted that there must always be sufficient fuel on hand to evacuate all personnel from the Antarctic in the event of any of the three aircraft becoming unserviceable.

The successful conclusion of ANI's first season of operating aircraft on wheels demonstrated the feasibility of carrying useful loads on direct flights from South America to high latitudes in the Antarctic. Now scientists and others could fly from South America and onward to their areas of interest at a fraction of the previous cost. For the first time, scientists from countries that do not have ships or bases in the Antarctic could work on the continent without either.

The anomalous position of Canadian citizens, operating a fleet of aircraft in the Antarctic without their country being a signatory of the Antarctic Treaty, was finally resolved on 4 May 1988 when the Government of Canada became the thirty-eighth state to accede to the treaty. They had shunned it for 29 years. I suspect that, in some measure, the change was a response to ANI's activities.

EPILOGUE

People involved in polar field work are often asked if it is risky. Their trite but sincere response is that it can be more dangerous to cross the road or drive a car at home. We learn to cope with Antarctic hazards as others learn to survive busy highways. The secret of survival in either environment is awareness. We take risks, just as people take risks on busy highways. We would not achieve much in the Antarctic without taking risks. However, we go to immense lengths to reduce risks to the minimum that is consistent with achieving our objective. The Norwegian explorer Roald Amundsen put it succinctly:

> Both imagination and caution are equally necessary – imagination to foresee the difficulties, and the caution which compels the minutest preparations to meet them.

Then in a cynical but understandable comment on the response of his detractors to his attainment of the South Pole:

> Victory awaits him who has everything in order – people call that luck. Defeat is certain for him who has neglected to take the necessary precautions in time – this is called bad luck.[1]

Travelling within the Antarctic has become safer because we make more use of aircraft. Admiral Byrd, with his Norwegian pilot Bernt Balchen, was the great pioneer in the wider use of aircraft. In relation to the perceived hazards, Antarctic aviation has on the whole a good safety record. Although we learn to cope with whiteout, blizzards, ice in fuel, and sometimes only sketch maps to navigate with, at the same time we are blessed with the world's largest potential landing ground for ski-equipped aircraft – 13 million square kilometres of it. Garry Studd once suffered a failure of both engines on his Twin Otter, and on another occasion a US Navy LC-130 lost power on three of its four engines. Both aircraft landed safely and, after clearing ice from fuel tanks, took off again.[2] Where else in the world could there have been such happy endings? At home the pilot in a similar emergency would be fighting to avoid buildings, earthworks, trucks, trees, cattle and a host of other obstructions.

Accidents will happen from time to time – we can never avoid all risks. Particularly in my own profession of glaciology, all too many of my friends have set out and never returned. However, if I could be in touch with them beyond the grave, I believe that not one would regret the career he chose.

***

One of the many changes that I have seen has been the coming of women to the Antarctic. The battle for equality took many years. Antarctica had always been regarded as a man's province. Indeed I have met some men who came south to escape from women. Wally Herbert had published a book under the title *A World of Men*.[3]

Two members of the Ronne Antarctic Research Expedition of 1947–48 had been women. Finn Ronne's wife Edith and Harry Darlington's wife Jenny spent the winter of 1947 at Stonington Island.[4] However, that was on a private expedition. Governments are more conservative. The long struggle has been documented by Elizabeth Chipman.[5] Her book *Women on the Ice* is, among some Antarcticans, unchivalrously referred to as 'Frigid Women'.

The French Antarctic leader Paul-Émile Victor was quoted as saying: 'We already have enough worries and I see no reason why we should help to create new ones.'[6] Yet history records that men's attitudes have not always been so negative. Robert Edwin Peary (1856–1920), who spent his life trying to reach the North Pole, wrote:

> Feminine companionship not only causes greater contentment, but as a matter of both mental and physical health and the retention of the top notch of manhood is a necessity.[7]

Peary underscored his point by taking an Eskimo 'wife' in Greenland (though he was already married) and fathering her child.

Paul Siple, after wintering at the South Pole in 1957, wrote: 'We agreed that what the Antarctic lacked most were women.'[8] The Russians, however, have had women working in the Antarctic during summer seasons since 1957. On the American side, it was not until 1969 that the first woman was allowed to work at McMurdo and in the field. One Antarctic veteran commented: 'The only place left now is the Moon!'[9]

Peary's reasons for wanting female company would be unfashionable today. BAS openly discriminated against women until quite recently. The logjam was broken by Janet Thomson, a geologist, who accepted an invitation from Ian Dalziel in 1976 to join an American ship-borne reconnaissance in the Antarctic Peninsula area. However, it was not until the 1983–84 season that she was able to take part in one of BAS's own ship-borne operations. Camping in tents was still considered unacceptable.

The Chilean Presidente Frei station on King George Island, with part of the Russian Bellingshausen station (far right). The small white buildings (left) are for officers and their families. The hangar at Marsh airstrip is on the horizon to the left of the orange fuel tank

Antarctic Airways DC-4 unloading fuel at Marsh. The tail in the foreground belongs to one of ANI's Twin Otters

First DC-4 landing at Patriot Hills. The red down suits are worn by Chileans

An engineer climbed out on the wing in a 35-knot wind at –25°C – then got sprayed with avgas

The camp at Patriot Hills. One of the tents blew away

In camp during the only nearly-calm day

Adventure Network's inaugural Hercules flight to Patriot Hills, 21 November 1993: General Javier Lopetegui, Hilda Reimer, Rachel Shephard and the author (standing)

Polar Logistics comes to Dronning Maud Land, 13 December 1996. The Hercules (right) has just landed after a 9-hour, 4,230 km flight from Cape Town. The Twin Otter (left) carries passengers and cargo from the ice runway to destinations within 1,000 km. Beyond the Hercules lies the peak of Ulvetanna – the Wolf's Fang (2,931 m). The surface in the foreground is typical of the 5-cm scale roughness of Antarctic ice runways

The Americans played a part in the next stage of the story. Janet was invited by Dr Peter Rowley of the US Geological Survey to join a field party working in Ellsworth Land during the 1984–85 season. By that time the Americans were getting used to having women in field camps but for BAS it was a milestone. Janet accepted, but could not share a tent.

BAS continued to prohibit any woman from flying in a BAS aircraft alone with one man (the pilot) on the grounds that in an emergency, the pair would have to bivouac without a chaperone. Since all field parties had to be flown into their work area, this meant that women could not take part.

It was not until 1988 that there was a change in policy. The first to benefit was Liz Morris, Head of the Ice and Climate Division, who worked with two men on George VI Ice Shelf in 1988–89, sharing a tent with each of them in turn.

Women were still excluded from wintering parties. That final barrier was broken when two women spent the winter of 1993 at the BAS station on Signy Island.

***

I look back on my 40 years on ice with enormous satisfaction, aware of what an incalculable privilege it has been to work with two generations of polar trailblazers. Science is one of the greatest adventures known to man, though scientists seem to shun the word because of its association with people who seek adventure for its own sake. However, no scientist can work without a raft of supporters who get their satisfaction not from any discovery but from the adventure itself. Luckily we work well together.

I was fortunate to hold a research post at the Scott Polar Research Institute for 12 years, and in retirement, I am still closely connected. Though much smaller than BAS, the institute has an enviable reputation for the scope and character of its research. It also has a postgraduate teaching programme towards a Master's Degree in Polar Studies. Being the only course of its kind, and covering both north and south polar regions, it draws students from nearly every country involved in polar research. Ph.D. degrees have been awarded in geology, physics, geophysics, anthropology, ethnology, politics and other subjects – all of them related to the polar regions. SPRI has probably produced more Ph.D. graduates in glaciology than has any other UK university. It has a large and well-organized polar library and for that reason alone can attract scholars from all over the world. Unlike BAS, it owns neither ships nor aircraft, relying instead on collaborative projects with larger organizations. This policy has succeeded, and can only succeed because members of the research staff are held in high esteem internationally. Many staff members of the institute have undertaken fieldwork in the Arctic, the Antarctic, or both.

The British Antarctic Survey, in contrast, is a massive enterprise. One objective of my own glaciological work with BAS was to establish a base-line against which future changes in the Antarctic ice sheet could be measured. Forty years after I first went to the Antarctic, the known changes are far greater than anything that I could have imagined. Thousands of square kilometres of the ice shelves round the Antarctic Peninsula have vanished.[10] In this remote corner of the world, we have seen a major change in climate over the last 50 years.[11]

The importance of this kind of work is that the Antarctic ice sheet serves as the principal control on sea level throughout the world. One good reason for trying to understand the mechanism of change is that – excluding water stored underground – 99 per cent of the fresh water on the surface of the earth is in the form of ice.[12] Contemporary changes are not confined to the Antarctic Peninsula. Globally averaged mean temperatures have shown a net warming over the last 130 years during which time sea level has risen.[13] Though there are compensating mechanisms (negative feedbacks) in the climatic system, we are learning to expect the unexpected. There is a growing realization that mankind, by altering the composition of the atmosphere, is changing the face of the earth.[14] Changes in the amount of ice will prove a crucial part of the story.

The work with which I was involved covers only a very small part of BAS research. No autobiographical account can do justice to the many other activities going on at the same time. BAS is a highly professional and uniquely productive research enterprise of which British science is justifiably proud.[15] Internationally, it probably leads the world both in the quality and also the cost-effectiveness of its work. At the same time it is staffed by a band of enthusiasts. These attributes are, no doubt, related.

The ten years that have elapsed since my retirement gave me the opportunity to play a part in improving access to the interior of Antarctica. Adventure Network International, the company that sponsored my initial contributions to the work, has confounded many a prophet of doom by maintaining a safe record for 12 years. The recipe was simple – employing people who, by virtue of previous experience, were up to the task. In terms of aircraft, the DC-4 yielded to a pressurized DC-6B in 1989 and to a much larger and more modern Lockheed L-382G Hercules four years later. I was aboard the first Hercules flight – in November 1993. A second Hercules was used in December 1996 to open a new air route from South Africa to Dronning Maud Land – I was aboard that too.

I accept that because of the harsh environment, Antarctic aviation may never match the safety record of the world's major airlines. However, ANI has matched the record of government operators.

Confounding predictions that Coca-Cola cans and cigarette ends would litter the wilderness, ANI has operated the cleanest camp in Antarctica and the *only* station from which all rubbish – including human waste – is taken out of the Antarctic. ANI was the first commercial operation in the Antarctic to commission an independent environmental audit of its activities in accordance with the requirements of the 1991 Protocol in Environmental Protection to the Antarctic Treaty.[16] Though ANI was set up for adventurers, it has also carried a number of government expeditions. Aware that national operators might balk at dealing with an 'adventure' company, ANI established an independent arm, Polar Logistics, to negotiate government commissions. There are indications that government-supported clients could soon outnumber the adventurers. The company is, after all, the only commercial airline operating to the mainland of Antarctica.

Blue icefields have been around since man first walked the earth. The concept of using them has been around for 20 years.[17] What was new was finding people with the courage to accept the risks involved in proving the concept. Sadly for aviation in general, Giles Kershaw, who led the way, died in a flying accident in 1990.[18] The accident was unrelated to his work with ANI.

In 1986 I predicted that government activities in Antarctica would have as much to gain from the development of airfields on ice as anyone else. After ANI had led the way, I was invited to spend six weeks during the 1988–89 season searching for new blue ice runways on behalf of the US Antarctic Program. Based on the South Pole, I was able to reconnoitre 30 blue-ice sites with a Twin Otter. As a direct result, National Science Foundation LC-130 aircraft have made wheel-landings on an icefield on Mill Glacier since 1990 and at Patriot Hills since 1993. Fuerza Aérea de Chile C-130 aircraft have landed at Patriot Hills since 1995; and a British Antarctic Survey DHC-7 landed on a blue icefield on the Palmer Land plateau in 1995. Other operators will follow suit when the need arises. In due course, I have no doubt that jet aircraft will land on inland icefields.

***

I have heard it said that today, with easier access to high latitudes, the challenge has gone out of polar fieldwork. Those who believe it should stay at home – to avoid being disillusioned.

NOTES

Prologue

1. John Giaever, *The White Desert: The official account of the Norwegian-British-Swedish Antarctic Expedition,* (London, Chatto and Windus, 1954)
2. Contemporary usage is in one stage of a slow transition from imperial to metric units. In a scientific paper I would adhere uncompromisingly to the metric system. However, this narrative is not addressed to scientists, so I found that consistency was unhelpful. The speed of ships and aircraft would be reported in metres per second, whereas knots (nautical miles per hour) are hallowed by usage and enshrined in international conventions. My compromises are as follows: speeds are given in knots, distances in kilometres, and heights in metres – except in quotations. Weights and the displacement of ships are in metric tonnes. When taken from American sources using long tons (1.016 tonnes), I have converted to metric tonnes. However, rounding of quantities makes some distinctions negligible.
3. Charles Swithinbank. *Norwegian-British-Swedish Antarctic Expedition 1949–52, Scientific Results,* 3, (Oslo, Norsk Polarinstitutt, 1957–60)

Chapter 1

1. Geoffrey Hattersley-Smith, 'Obituary: Trevor Arnold Harwood', *Polar Record*, 22:140 (1985), pp. 555–6
2. Moira Dunbar and Keith R. Greenaway, *Arctic Canada from the Air,* (Canada, Defence Research Board, 1956)
3. G. Hattersley-Smith, *North of Latitude Eighty, the Defence Research Board in Ellesmere Island,* (Canada, Defence Research Board, 1974)
4. The concentration of pack ice is generally reported in terms of tenths of the sea surface covered by ice floes. Very open pack ice covers 1/10th to 3/10ths; open pack ice 4/10ths to 6/10ths; close pack ice 7/10ths to 9/10ths; very close pack ice almost 10/10ths; and consolidated pack ice 10/10ths with the floes frozen together.
5. David Maclellan, 'Pullen of the Arctic', *Canadian Geographic*, 104:2 (1984), pp.26–33

6. An exhaustive discourse on the expedition is: Richard Julius Cyriax, *Sir John Franklin's Last Expedition; a chapter in the history of the Royal Navy*, (London, Methuen, 1939). An easier read is: Roderic Owen, *The Fate of Franklin*, (London, Hutchinson, 1978)

7. Everyone I met in 1956 spoke of Eskimos. The preferred term today is Inuit, though some Inuit may refer to themselves as Eskimos.

8. The word *station* only takes a capital *S* when it is part of a proper name.

9. Graham Rowley, 'Captain T.C. Pullen, RCN: Polar Navigator', *The Northern Mariner*, 2:2 (1992), pp. 29–49

10. The naval term is *screw*. In this narrative I discuss naval and civil ships – sometimes in the same convoy – so have chosen to use *propeller* throughout.

11. Apart from concentration, the main terms used in describing the development of pack ice are new ice, first-year ice, second-year ice, or multi-year ice. Ice floes may decay to become rotten ice or brash ice.

12. Henry Asbjorn Larsen, *The Northwest Passage 1940–42 and 1944: the famous voyages of the Royal Canadian Mounted Police Schooner St. Roch,* (Ottawa, Queen's Printer, 1958)

13. Thirty-two years later, in 1988, I revisited Fort Ross in the course of a voyage – once again with Captain Pullen – in the polar cruise ship *Society Explorer*. The house was still in good condition.

14. Walter Kowal, Owen B. Beattie, Halfdan Baadsgaard and Peter M. Krahn, 'Source identification of lead in tissues of sailors from the Franklin Arctic Expedition of 1845', *Journal of Archaeological Science*, 18 (1991), pp. 193–203

15. In four years of working on this project I travelled 100,000 kilometres and visited Dansk Meteorologisk Institut, Charlottenlund, Denmark; Arktisk Institut, Copenhagen; Universitetsbiblioteket, Oslo; Norsk Polarinstitutt, Oslo; Royal Geographical Society, London; British Museum (Natural History), London; Meteorological Office, Air Ministry, Harrow, England; Museum of Fisheries and Whaling, Hull; National Maritime Museum, Greenwich; Hudson's Bay Company, London; Hudson's Bay Company, Winnipeg, Manitoba; Arctic Institute of North America, Ottawa; Arctic Institute of North America, Montreal, Quebec;. Provincial Archives, Halifax, Nova Scotia; Geographical Branch, Department of Mines and Technical Surveys, Ottawa; Canadian Hydrographic Service, Department of Mines and Technical Surveys, Ottawa; Public Archives of Canada, Ottawa; National Air Photo Library, Ottawa; Department of Transport, Ottawa; Geological Survey of Canada, Ottawa; Department of Northern Affairs and National Resources, Ottawa; US Coast and Geodetic Survey,

Washington, DC; US Coast Guard Headquarters, Washington, DC; US National Archives, Washington, DC; US Navy Hydrographic Office, Suitland, Maryland; US Weather Bureau, Washington, DC; Library of Congress, Washington, DC; Arctic Institute of North America, New York; American Geological Society, New York; Old Dartmouth Historical Society and Whaling Museum, New Bedford, Massachusetts; Free Public Library, New Bedford, Massachusetts; Woods Hole Oceanographic Institution, Woods Hole, Massachusetts; Nantucket Whaling Museum, Nantucket, Massachusetts; Stefansson Collection, Baker Library, Dartmouth College, Hanover, New Hampshire; US Army, Snow, Ice, and Permafrost Research Establishment, Wilmette, Illinois. Two hundred and fifty ships contributed ice data from more than 600 voyages; 50,000 aerial photographs were examined on behalf of the project by Diana Rowley; and published and unpublished ice charts from many sources were incorporated.

16. Graham Rowley described some of his own Arctic travels in *Cold Comfort: My love affair with the Arctic,* (Montreal, McGill-Queen's University Press, 1996)

17. Charles Swithinbank, *Ice Atlas of Arctic Canada,* (Ottawa, The Queen's Printer, 1960)

Chapter 2

1. Summer in the Antarctic extends November to February, thus from one calendar year to the next. Chapter headings bracketed with a single year refer to Arctic summers.

2. Gilbert Dewart, *Antarctic Comrades* (Columbus, The Ohio State University Press, 1989)

3. Some countries refer to stations, others to bases. The terms are interchangeable. When BAS began in 1943 it referred only to bases, but many people now use the alternative to avoid association with military bases.

4. M.E.R. Walford, 'Radio echo sounding through an ice shelf', *Nature*, 204:4956 (1964), pp. 317–19

5. S. Evans and G. de Q. Robin, 'Glacier depth sounding from the air', *Nature*, 210:5039 (1966), pp. 883–5

6. V.V. Bogorodsky, C.R. Bentley and P.E. Gudmandsen, *Radioglaciology*, (Boston, D. Reidel Publishing Company, 1985)

7. John Biscoe was a British sealer who circumnavigated Antarctica in the brig *Tula* in 1830–32.

8. In an Antarctic context, landing grounds on snow are described as airstrips, runways, or skiways. The terms are interchangeable. Airfields on rock, of which at this time there were only three on the whole conti-

nent, are variously described as airstrips or runways.

Chapter 3

1. Sir Ernest Shackleton, *South, the story of Shackleton's last expedition 1914–1917* (London, William Heinemann, 1919)
2. Aviation turbine fuel is also known as jet fuel, JP-1 or JP-4.
3. Charles Swithinbank, 'Radio echo sounding of Antarctic glaciers from light aircraft', *International Association of Scientific Hydrology*, Publication no. 79 (1968), pp. 405–14
4. D.J. Drewry (ed.), *Antarctica: glaciological and geophysical folio* (Cambridge, Scott Polar Research Institute, 1983)
5. Ayers to Director, 26 March 1967 (X/183/67)

Chapter 4

1. Bern Keating, *The Northwest Passage from the Mathew to the Manhattan: 1497 to 1969,* (Chicago, Rand McNally, 1970)
2. *BP Pacer*, September 1969
3. Robert L. Scheina, *US Coast Guard Cutters and Craft of World War II* (Annapolis, Maryland, Naval Institute Press) p. 56
4. *The New York Times*, 29 September 1969, p. C25

Chapter 5

1. Up to this time we had only rammed *ice floes* up to 3 metres thick. In contrast, *icebergs* can be 10–30 metres thick and consist of fresh-water ice, which is harder than sea ice.
2. T.C. Pullen, interviewed in *US News & World Report*, 9 February 1970, p. 74
3. A.D. Mookhoek and W.J. Bielstein, 'Problems associated with the design of an Arctic marine transportation system', *Third Annual Offshore Technology Conference, Houston, Texas, 19–21 April 1971*, 2, pp. 123–46
4. Roald Amundsen. *The North West Passage* (2 vols., London, Constable, 1908)
5. *The Globe and Mail*, Toronto, 12 August 1970
6. Humble Oil and Refining Company press release, 16 April 1971.
7. Charles Swithinbank, *SS Manhattan. Notes on a voyage in the Northwest Passage 1969* (proprietary)
8. David Baker and Roy Nishizaki, *The MV Arctic – new bow form and model testing,* (The Society of Naval Architects and Marine Engineers, 1986, unpublished)
9. *Society of Naval Architects and Marine Engineers Transactions*, 67

(1959), p. 184

10. *Financial Times*, 19 September 1975, p. 16
11. Cynthia Lamson and David L. Vanderzwaag (eds.), *Transit Management in the Northwest Passage: Problems and Prospects* (Cambridge University Press, 1988)
12. Humble Oil and Refining Company press release, 21 October 1970.
13. Vilhjalmur Stefansson, *The Northward Course of Empire* (New York, Harcourt Brace, 1922), pp. 189–99
14. Joel E. Truitt, 'Moving Alaska's oil by submarine tanker', *Navy*, 75:3 (March 1970), pp. 80–1

Chapter 6

1. Royal Navy submariners tend to call their vessel a *boat*, whereas *ship* is used by submariners elsewhere. When I was in the navy in 1944 my CO insisted that a *boat* was a craft that could be lifted out of the water onto a *ship*. Early submarines could, but not *Dreadnought*.
2. *Dreadnought's* officers were:
Cdr Alan G. Kennedy, Commanding Officer
Lt Cdr Robin C. Whiteside, First Lieutenant
Lt Cdr David M. Pulvertaft, Senior Engineering Officer
Lt Cdr D. Andrew McVean, Weapons Eng. Electrical Officer
Lt Cdr John H. Collier, Torpedo Anti-Submarine Officer
Lt Christopher L. Napier, Navigator
Lt R.F. Strange, Communicator and Naval Stores Officer
Lt Peter Jordan, Trimming Officer, Secretary
Lt Peter H. Aldous, Torpedo Officer
Lt Martin E. Murray, Asst. Mech. Engineer Officer
Lt R.M.P. Manson, Asst. Weapons Electrical Officer
Lt Christopher Field, Electrical Officer under training
Surgeon Lt F.M. John Hiles, Medical Officer

Chapter 7

1. William R. Anderson, *Nautilus 90 North* (London, Hodder and Stoughton, 1959), p. 156
2. James Calvert, *Surface at the Pole* (London, Hutchinson, 1961), p. 59
3. There have been many different calculations of wind-chill. See: W.G. Rees, 'A new wind-chill nomogram', *Polar Record*, 29:170 (1993), pp. 229–34
4. Twenty-three years later (in August 1994) I steamed across these latitudes in the Russian nuclear-powered icebreaker *Yamal* (75,000 shaft horsepower). The ship ploughed through every pressure ridge we

encountered without stopping. Very few of her crew grasped that we were simultaneously ploughing through ice keels 20–30 metres deep.

5. Charles Swithinbank, 'Arctic pack ice from below', *Sea Ice, Proceedings of an International Conference, Reykjavik, Iceland, May 10–13, 1971* (Reykjavik, National Research Council, 1972), pp. 246–54

6. Elizabeth Williams, Charles Swithinbank and G. de Q. Robin, 'A submarine sonar study of Arctic pack ice', *Journal of Glaciology*, 15:73 (1975), pp. 349–62

Chapter 8

1. The International Civil Aviation Organization reports altitudes in *feet*, whereas scientists prefer *metres*. In order to avoid alternating according to context, I have chosen to stick to the metric system. The one exception is in quoted speech, where I report the units as spoken.

Chapter 9

1. M.E.R. Walford, 'Radio echo sounding through an ice shelf', *Nature*, 204:4956, (1964), pp. 317–19

2. B.M. Ewen Smith, 'Airborne radio echo sounding of glaciers in the Antarctic Peninsula', *British Antarctic Survey Scientific Reports*, 72 (1972), p. 8

3. This was the first of many measurements made over a period of years. The results were published: John M. Reynolds, 'The distribution of mean annual temperatures in the Antarctic Peninsula', *British Antarctic Survey Bulletin*, 54 (1981), pp. 123–33

4. In 1975 it was named Gipps Ice Rise after Derek Gipps, a long-serving BAS Senior Executive Officer.

Chapter 10

1. Sir Vivian Fuchs, *The Crossing of Antarctica. The Commonwealth Trans-Antarctic Expedition 1955–58,* (London, Cassell, 1958), pp. 111–12

2. M.E.R. Walford, 'Glacier movement measured with a radio echo technique', *Nature*, 239:5367 (1972), pp. 95–6

3. Charles Swithinbank, 'A new map of Alexander Island, Antarctica', *Polar Record*, 17:107 (1974), pp. 155–7

4. Nevertheless, we thought it wise to repeat the measurements over a longer period. I arranged for Christopher Doake to do this two years later. He repeated the experiment and found that the discrepancy between the two methods remained. Chris concluded that the radio-echo system was measuring the surface velocity in relation to rocks being carried along at the base of the ice – it was not seeing the solid bedrock

beneath. The difference between his results and the conventional survey results therefore represented the sliding velocity of the glacier over its bed. This was not what we had set out to measure but it was possibly more interesting because nobody had ever devised a way to measure bottom sliding without drilling to the bed of a glacier (C.S.M. Doake, 'Glacier sliding measured by a radio-echo technique', *Journal of Glaciology*, 15:73 (1974), pp. 89–93

Chapter 11

1. Some years later I arranged for a consultant on airfield construction to examine the site. His report confirmed our findings but he too recognized the iceberg hazard (Kenneth M. Adam, *Assessment of the proposed Rothera Point airstrip, Adelaide Island, Antarctica*. Winnipeg, I.D. Engineering Company, 1983-unpublished). It took 16 years to raise sufficient interest and finance. But a fine runway, longer and wider than we had proposed, was finally opened in 1991. It is one of only three hard runways in the whole of Antarctica. The others are at Marsh (Chile) on King George Island and Marambio (Argentina) on Seymour Island. BAS aircraft have operated safely from the airstrip ever since. However, the consultant's note about the iceberg hazard proved prophetic. In 1994 a visiting Canadian Twin Otter crashed into an iceberg on take-off, killing all on board (*AAIB Bulletin 2/95*. Farnborough, Air Accidents Investigation Branch, 1995).
2. This was Bert Conchie's last Antarctic flying season. By March 1975 he had flown in six consecutive summer seasons.
3. This was Rutford Glacier, later renamed Rutford Ice Stream. Robert Hoxie Rutford was Director of the Division of Polar Programs at the US National Science Foundation, 1975–77.
4. Small single-tracked vehicles used in Antarctica have been variously described as motor sledges, motor toboggans, snow-scooters, snowmobiles and skidoos. The terms are interchangeable.

Chapter 12

1. K.J. Bertrand, *Americans in Antarctica 1775–1948* (New York, American Geographical Society, 1971).
2. In 1977 this arm was named Evans Ice Stream after Dr Stanley Evans, designer of the Scott Polar Research Institute radio-echo sounder.
3. Named Fowler Ice Rise by the Americans in 1976 and Fowler Peninsula by the British in 1980 on the basis of our findings. Captain Alfred N. Fowler, USN, was Commander of the US Naval Support Force, Antarctica, 1972–74.
4. Named Carlson Inlet in 1976 after Lieutenant (later Commander)

Ronald F. Carlson, USN, the R4D pilot who in 1961 flew me over the Ross Ice Shelf and later searched for my party during a radio blackout.

5. Named Fletcher Ice Rise in 1976 by the Americans and Fletcher Promontory in 1980 by the British on the basis of our findings. Dr Joseph Otis Fletcher was Director of the US National Science Foundation Office of Polar Programs, 1971–74.

6. This 'locally grounded area' was named Kershaw Ice Rumples in 1977. He deserved it.

7. Named Korff Island in 1958 but renamed Korff Ice Rise in 1977. Serge Alexander Korff was an American of Finnish birth, a physicist and cosmic ray specialist.

8. Named Zumberge Coast in 1976. James Herbert Zumberge, geologist (1924–92), President of Grand Valley State College, and later, in succession, Dean of the College of Earth Sciences at the University of Arizona, Chancellor of the University of Nebraska–Lincoln, President of Southern Methodist University, and finally President of the University of Southern California at Los Angeles.

9. Charles Swithinbank, Christopher Doake, Andrew Wager and Richard Crabtree, 'Major change in the map of Antarctica', *Polar Record*, 18:114 (1976), pp. 295–9

10. Charles Swithinbank, 'Glaciological Research in the Antarctic Peninsula', *Philosophical Transactions of the Royal Society*, B 279, (1977), pp. 161–83

11. Finn Ronne, 'Ronne Antarctic Research Expedition 1946–48', *Geographical Review*, 38:3 (1948), pp. 355–91

12. P.D. Clarkson and Maureen Brook, 'Age and position of the Ellsworth Mountains crustal fragment, Antarctica', *Nature*, 265:55 (1977), pp. 615–16

Chapter 13

1. Captain Sir James Clark Ross, *A Voyage of Discovery and Research in the Southern and Antarctic Regions During the Years 1839–43* (2 vols., London, John Murray, 1847), Vol. 2, p. 332

2. Charles Swithinbank, 'Satellite image atlas of glaciers of the world: Antarctica', *US Geological Survey Professional Paper 1386-B* (1988), p. 112

3. *British Antarctic Territory Ice Thickness Map 1:500,000*, Series BAS 500 R, Sheet 1, 1983

4. *Filchner-Ronne Ice Shelf Programme – Reports* (Bremerhaven, Alfred-Wegener-Institute for Polar Research). Published annually from no. 1 (1984) to no. 10 (1996)

5. C.S.M. Doake, R.M. Frolich, D.R. Mantripp, A.M. Smith and D.G.

Vaughan, 'Glaciological studies on Rutford Ice Stream, Antarctica', *Journal of Geophysical Research*, 92 (B9) (1987), pp. 8951–60
6. K.W. Nicholls, K. Makinson and V.A. Robinson, 'Ocean circulation beneath Ronne Ice Shelf'. *Nature*, 354: 6350 (1991), pp. 221–3
7. Charles Swithinbank, 'Satellite image atlas of glaciers of the world: Antarctica', *US Geological Survey Professional Paper* 1386-B (1988)
8. Malcolm Mellor and Charles Swithinbank, 'Airfields on Antarctic glacier ice', *CRREL Report* 89-21 (Hanover, New Hampshire, US Army Cold Regions Research and Engineering Laboratory, 1989)
9. This commendable initiative was overtaken in April 1982 when Argentine forces took the Falkland Islands by storm. Although British forces recovered the islands ten weeks later, relations between the two countries were set back for a generation.
10. C.W.M. Swithinbank and C. Lane, 'Antarctic Mapping from Satellite Imagery', *Conference of Commonwealth Survey Officers 1975*, Chapter 11, (London, Ministry of Overseas Development, 1975)
11. An earthquake in 1985 left 40,000 homeless.

Chapter 14

1. Adelaide station was closed on 1 March 1977, having been occupied continuously for 16 years. Rothera station took over responsibility for aircraft operations.
2. Ranulph Fiennes, *To the Ends of the Earth* (London, Hodder and Stoughton, 1983)
3. Ten years later Geoff Somers represented Britain on an international expedition that crossed Antarctica the long way – from Graham Land to Mirny. See: Will Steger and Jon Bowermaster, *Crossing Antarctica,* (London, Bantam Press, 1992)
4. Alan Reece, 'The ice of Crown Prince Gustav Channel, Graham Land, Antarctica', *Journal of Glaciology*, 1:8 (1950), pp. 404–9

Chapter 15

1. Robert Headland, *The Island of South Georgia* (Cambridge University Press, 1984), p. 252

Chapter 16

1. The fuel system was dismantled after the aircraft landed in Chile. One main fuel tank was found to contain 2 kg of sugar.
2. Javier Lopetegui Torres, *Antártica – un Desafío Perentorio* (Santiago, Instituto Geopolitico de Chile, 1986)
3. Dick Bass, Frank Wells and Rick Ridgeway, *Seven Summits* (New York, Warner Books, 1986)

4. The airstrip is now known by the name of its adjacent coastal station – Presidente Eduardo Frei.
5. None of these factors deterred them. The attraction of a ready-made base was too hard to resist. After government-to-government discussions, the Chileans were offered the station at no cost. The transfer gave rise to a question in the House of Commons about the possible implications for sovereignty. However, under the terms of Article IV of the Antarctic Treaty, there were no such implications. The Chileans have used Adelaide ever since as a summer weather station and staging post. They renamed it Teniente Luis Carvajal.

Chapter 17

1. In the event, there was not only a second but also a third season. The principal results were published in:
B.C. Storey, I.W.D. Dalziel, S.W. Garrett, A.M. Grunow, R.J. Pankhurst and W.R. Vennum, 'West Antarctica in Gondwanaland: Crustal blocks, reconstruction, and breakup processes'. *Tectonophysics*, 155 (1988), pp. 381–90
2. Letter of 29 August 1983 from Grikurov to Swithinbank.
3. On 24 February 1985 another Dornier 228 aircraft home-bound from the West German Antarctic station Georg von Neumayer was shot down over the Western Sahara by Polisario guerillas who had mistaken it for a Moroccan aircraft. All on board were killed. It was the first war casualty in the history of Antarctic aviation.

Chapter 18

1. Stuart Macfarlane, *The Erebus Papers* (Auckland, Avon Press, 1991)
2. Ranulph Fiennes, *To the Ends of the Earth* (London, Hodder and Stoughton, 1983)
3. Valter Schytt, 'Blue ice-fields, moraine features and glacier fluctuations', *Norwegian-British-Swedish Antarctic Expedition 1949–52, Scientific Results*, 4E (Oslo, Norsk Polarinstitutt, 1960)
4. Karsten Brunk and Rudolf Staiger, 'Nachmessungen an Pegeln auf einem Blaueisfeld im Borgmassiv, Neuschwabenland, Antarktis', *Polarforschung*, 56:1–2, pp. 23–32
5. Charles Swithinbank, Ice runways in the Heritage Range, Antarctica. A report prepared for Adventure Network International and Fuerza Aérea de Chile (Cambridge, February 1987-unpublished)

Chapter 19

1. Samuel C. Silverstein, 'The American Antarctic Mountaineering

Expedition', *Antarctic Journal of the United States*, 2:2 (1967), pp. 48–50

2. Charles Swithinbank, 'Airborne tourism in the Antarctic', *Polar Record*, 29:169 (1993), pp. 103–10

Epilogue

1. Roald Amundsen, *My Life as an Explorer* (London, Heinemann, 1927), pp. 258–9
2. David M. Tyree, 'New era in the loneliest continent', *National Geographic Magazine*, 123:2 (1963), p. 283
3. Wally Herbert, *A World of Men* (London, Eyre and Spottiswoode, 1968)
4. Jennie Darlington, *My Antarctic Honeymoon: a year at the bottom of the world* (Garden City, NY, Doubleday, 1956)
5. Elizabeth Chipman, *Women on the Ice. A history of women in the far south* (Melbourne University Press, 1986)
6. *Antarctic* (a news bulletin published quarterly by the New Zealand Antarctic Society), 4:1 (1965), p. 14. See also: Paul-Émile Victor, *Mes aventures polaires* (Paris, Éditions G.P., 1975). M. Victor died on 7 March 1995, aged 87.
7. Wally Herbert, *The Noose of Laurels* (London, Hodder and Stoughton, 1989), p. 62
8. Paul Siple, *90° South. The story of the American South Pole conquest,* (New York, G.P. Putnam's Sons, 1959), p. 338
9. Walter Sullivan, in *The New York Times*, 1 October 1969. Quoted in Chipman, *op. cit.*, p. 95
10. D.G. Vaughan and C.S.M. Doake, 'Recent atmospheric warming and retreat of ice shelves on the Antarctic Peninsula', *Nature*, 379:6563 (1996), p. 328–31
    Helmut Rott, Pedro Skvarca and Thomas Nagler, 'Rapid collapse of northern Larsen Ice Shelf, Antarctica', *Science*, 271:5250 (1996), p. 788–92
11. John C. King, 'Recent climatic variability in the vicinity of the Antarctic Peninsula', *International Journal of Climatology*, 14:4 (1994), pp. 357–69
12. H. Hoinkes, 'Glaciology in the International Hydrological Decade', *International Association of Scientific Hydrology*, 79 (1968), p. 7
13. J.T. Houghton, L.G. Meiro Filho, B.A. Callander, N. Harris, A. Kattenburg and K. Maskell (eds.), *Climate Change 1995. The science of climatic change. Contribution of WGI to the second assessment report of the Intergovernmental Panel on Climate Change.* (Cambridge University Press, 1996)
14. T.M.L. Wigley, R. Richels and J.A. Edmonds, 'Economic and environ-

mental choices in the stabilization of atmospheric CO_2 concentrations', *Nature*, 379:6562 (1996), pp. 240–3

15. Readers interested in a fuller account should refer to the *British Antarctic Survey Annual Report* series published by the Natural Environment Research Council
16. *Initial Environmental Evaluation: Adventure Network International, Antarctic Airborne Operations* (Cambridge, Poles Apart, 1995)
17. A. Kovacs and G. Abele, 'Runway Site Survey, Pensacola Mountains, Antarctica'. *CRREL Special Report 77–14,* (Hanover, New Hampshire, US Army Cold Regions Research and Engineering Laboratory, 1977)
18. Charles Swithinbank, 'Obituary: John Edward Giles Kershaw', *Polar Record*, 26:158 (1990), p. 250

ACRONYMS

ANI	Adventure Network International
BAS	British Antarctic Survey
CCGS	Canadian Coast Guard Ship
CRREL	Cold Regions Research and Engineering Laboratory of the US Army Corps of Engineers
DEW	Distant Early Warning
ETA	Estimated Time of Arrival
FACH	Fuerza Aérea de Chile (Chilean Air Force)
FID or **FIDS**	Members of the Falkland Islands Dependencies Survey
HELO	Helicopter
HMS	Her Majesty's Ship
HMCS	Her Majesty's Canadian Ship
INS	Inertial Navigation System
LCVP	Landing Craft Vehicle and Personnel
LSD	Landing Ship Dock
NERC	Natural Environment Research Council
NSF	National Science Foundation (Washington, DC)
RAF	Royal Air Force
RCAF	Royal Canadian Air Force
RCMP	Royal Canadian Mounted Police
RCN	Royal Canadian Navy
RN	Royal Navy
RRS	Royal Research Ship
SCAR	Scientific Committee on Antarctic Research (of the International Council of Scientific Unions)
SINS	Submarine Inertial Navigation System
SPRI	Scott Polar Research Institute
USAP	United States Antarctic Program
USCGC	United States Coast Guard Cutter
VIP	Very Important Person

GLOSSARY

Altimeter	Aneroid barometer used to measure altitude of aircraft.
Barrier	Obsolete name for ice front and/or ice shelf.
Bergy bit	A piece of floating ice, generally less than 5 m above sea level and not more than about 10 m across.
Caldera	Large scale volcanic crater often formed by subsidence.
Calving	The breaking away of a mass of ice from a floating ice shelf, glacier or iceberg.
Conglomerate	Rock composed of fragments of pre-existing rocks cemented together.
Crampon	Spiked metal frame strapped to boots to facilitate climbing on ice.
Crevasse	A fissure formed in a glacier, ice sheet or ice shelf. Crevasses are often hidden by snow bridges.
Depot	Supplies left in the field for later use.
Erratic	Rock transported from its original location, usually by the action of ice.
Fast ice	Sea ice which remains fast along the coast, where it is attached to the shore or to an ice shelf. An abbreviation of landfast ice.
Floe	A piece of floating ice other than fast ice or glacier ice.
Galley	Naval term for kitchen. In practice used to include dining room.
Hypothermia	Cooling of the body to danger level as a result of heat loss from exposure.
Iceberg	A large mass of floating or stranded ice which has broken away from a glacier.
Ice drill	A coring device for obtaining ice samples. May be hand- or motor-powered.
Ice edge	The boundary at any given time between pack ice and open water.
Icefall	A heavily crevassed area in a glacier at a region of steep descent.
Ice front	The vertical cliff forming the seaward face of an ice shelf or other floating glacier, varying in height from two to 40-metres above sea level.

Ice piedmont	Ice covering a coastal strip of low-lying land backed by mountains.
Ice sheet	A mass of ice and snow of considerable thickness and large area. Ice sheets may be resting on rock (see *Inland ice sheet*) or floating (see *Ice Shelf*). Ice sheets of less than 50,000 square kilometres in area are called *ice caps.*
Ice shelf	A floating ice sheet of considerable thickness attached to a coast. Ice shelves are usually of great horizontal extent and have a level or gently undulating surface. They are nourished by the accumulation of snow and often by the seaward extension of land glaciers. Limited areas may be aground. The seaward edge is termed an *ice front.*
Ice stream	Part of an ice sheet in which the ice flows more rapidly and not necessarily in the same direction as the surrounding ice. The margins are sometimes clearly marked by a change in direction of the surface slope, but may be indistinct.
Inland ice sheet	An ice sheet of considerable thickness and more than about 50,000 square kilometres in area, resting on rock. Inland ice sheets near sea level may merge into ice shelves.
Knot	A unit of speed equal to one nautical mile per hour.
Lead	A navigable passage through floating ice.
Moraine	Ridges or deposits of rock debris transported by a glacier. Common forms are: lateral moraine, along the sides; medial moraine, down the centre.
Nunatak	A rocky crag or small mountain projecting from and surrounded by a glacier or ice sheet.
Pack ice	An area of sea ice other than fast ice, no matter what form it takes or how it is disposed.
Sastrugi	Sharp, irregular ridges formed on a snow surface by wind erosion and deposition. The ridges are usually parallel to the prevailing wind.
Sea ice	Any form of ice found at sea which originated from the freezing of sea water.
Skidoo	An alternative word for snowmobile, motor sledge, or motor toboggan.
Sno-cat	An oversnow vehicle sprung over four separate steerable tracks.
Snow bridge	An arch formed by snow which has drifted across a crevasse, forming first a cornice, and ultimately a covering which may completely obscure the opening.

Snowdrift	An accumulation of wind-blown snow deposited in the lee of obstructions or heaped by wind eddies.
Snowmobile	Single-tracked vehicle with steerable ski at the front.
Strand crack	A fissure at the junction between an inland ice sheet, ice piedmont or ice rise, and an ice shelf, the latter being subject to the rise and fall of the tide.
Sublimation	Conversion of ice or snow to vapour without passing through the liquid phase.
Theodolite	A precise angle-measuring instrument consisting of a telescopic sight mounted on graduated horizontal and vertical circles.
Valley glacier	A glacier which flows down a valley.
Whiteout	A condition in which daylight is diffused by multiple reflection between a snow surface and overcast sky. Contrasts vanish and the observer is unable to distinguish snow surface features.

REFERENCES

AAIB Bulletin 2/95. Farnborough, Air Accidents Investigation Branch, 1995

Adam, Kenneth M., *Assessment of the proposed Rothera Point airstrip, Adelaide Island, Antarctica*, Winnipeg, I.D. Engineering Company, 1983 (unpublished)

Amundsen, Roald, *My Life as an Explorer*, London, Heinemann, 1927 (pp. 258–9)

Amundsen, Roald, *The North West Passage*, 2 vols., London, Constable, 1908

Anderson, William R., *Nautilus 90 North*, London, Hodder and Stoughton, 1959 (p. 156)

Antarctic (a news bulletin published quarterly by the New Zealand Antarctic Society), vol. 4, no. 1, 1965 (p. 14)

Baker, David and Roy Nishizaki, 'The MV Arctic – new bow form and model testing,' The Society of Naval Architects and Marine Engineers, 1986 (unpublished)

Bass, Dick, Frank Wells and Rick Ridgeway, *Seven Summits*, New York, Warner Books, 1986

Bertrand, K.J. *Americans in Antarctica 1775–1948*, New York, American Geographical Society, 1971

Bogorodsky, V.V., C.R. Bentley and P.E. Gudmandsen, *Radioglaciology*, Boston, D. Reidel Publishing Company, 1985

BP Pacer, September 1969

British Antarctic Survey Annual Report series, Swindon, Natural Environment Research Council

British Antarctic Territory Ice Thickness Map 1:500,000, Series BAS 500 R, Sheet 1, 1983

Brunk, Karsten and Rudolf Staiger, 'Nachmessungen an Pegeln auf einem Blaueisfeld im Borgmassiv, Neuschwabenland, Antarktis', *Polarforschung*, vol. 56, no. 1–2 (pp. 23–32)

Calvert, James, *Surface at the Pole*, London, Hutchinson, 1961 (p. 59)

Chipman, Elizabeth, *Women on the Ice. A history of women in the far south*, Melbourne University Press, 1986

Clarkson, P.D. and Maureen Brook, 'Age and position of the Ellsworth Mountains crustal fragment, Antarctica', *Nature*, vol. 265, no. 55, 1977 (pp. 615–16)

Cyriax, Richard Julius, *Sir John Franklin's Last Expedition: a chapter in the history of the Royal Navy*, London, Methuen, 1939

Darlington, Jennie, *My Antarctic Honeymoon: a year at the bottom of the world*, Garden City, NY, Doubleday, 1956

Dewart, Gilbert, *Antarctic Comrades*, Columbus, The Ohio State University Press, 1989

Doake, C.S.M., 'Glacier sliding measured by a radio-echo technique'. *Journal of Glaciology*, vol. 15, no. 73, 1975 (pp. 89–93)

Doake, C.S.M., R.M. Frolich, D.R. Mantripp, A.M. Smith and D.G. Vaughan, 'Glaciological studies on Rutford Ice Stream, Antarctica'. *Journal of Geophysical Research*, vol. 92 (B9), 1987, pp. 8951–60

Drewry, D.J. (ed.), *Antarctica: glaciological and geophysical folio*, Cambridge, Scott Polar Research Institute, 1983

Dunbar, Moira and Keith R. Greenaway, *Arctic Canada from the Air*, Canada, Defence Research Board, 1956

Evans, S. and G. de Q. Robin, 'Glacier depth sounding from the air', *Nature*, vol. 210, no. 5039, 1966 (pp.883–5)

Fiennes, Ranulph, *To the Ends of the Earth*, London, Hodder and Stoughton, 1983

Filchner-Ronne Ice Shelf Programme – Reports, Bremerhaven, Alfred-Wegener-Institute for Polar Research. Published annually from no. 1 (1984) to no. 10 (1996)

Financial Times, London, 19 September 1975 (p. 16)

Fuchs, Sir Vivian, *The Crossing of Antarctica. The Commonwealth Trans-Antarctic Expedition 1955–58*, London, Cassell, 1958 (pp. 111–12)

Giaever, John, *The White Desert. The official account of the Norwegian-British-Swedish Antarctic Expedition*, London, Chatto and Windus, 1954

Globe and Mail, Toronto, 12 August 1970

Hattersley-Smith, G., *North of Latitude Eighty, the Defence Research Board in Ellesmere Island*, Canada, Defence Research Board, 1974

Hattersley-Smith, Geoffrey, 'Obituary: Trevor Arnold Harwood', *Polar Record*, vol. 22, no. 140, 1985 (pp. 555–6)

Headland, Robert, *The Island of South Georgia*, Cambridge University Press, 1984 (p. 252)

Herbert, Wally, *The Noose of Laurels*, London, Hodder and Stoughton, 1989 (p. 62)

Herbert, Wally, *A World of Men*, London, Eyre and Spottiswoode, 1968

Hoinkes, H., 'Glaciology in the International Hydrological Decade', *International Association of Scientific Hydrology*, Publication No. 79, 1968 (p. 7)

Houghton, J.T., L.G. Meiro Filho, B.A. Callander, N. Harris, A. Kattenburg and K. Maskell (eds.), *Climate Change 1995. The science of climatic change. Contribution of WGI to the second assessment report of the Intergovernmental Panel on Climate Change.* Cambridge University Press, 1996

Initial Environmental Evaluation: Adventure Network International, Antarctic Airborne Operations, Cambridge, Poles Apart, 1995

Keating, Bern, *The Northwest Passage from the Mathew to the Manhattan: 1497 to 1969*, Chicago, Rand McNally, 1970

King, John C., 'Recent climatic variability in the vicinity of the Antarctic Peninsula', *International Journal of Climatology*, vol. 14, no. 4, 1994 (pp. 357–69)

Kovacs, A. and G. Abele, 'Runway Site Survey, Pensacola Mountains, Antarctica'. *CRREL Special Report* 77–14, Hanover, New Hampshire, US Army Cold Regions Research and Engineering Laboratory, 1977

Kowal, Walter, Owen B. Beattie, Halfdan Baadsgaard and Peter M. Krahn, 'Source identification of lead in tissues of sailors from the Franklin Arctic Expedition of 1845,' *Journal of Archaeological Science*, vol. 18, 1991 (pp. 193–203)

Lamson, Cynthia and David L. Vanderzwaag (eds.), *Transit Management in the Northwest Passage: problems and prospects*, Cambridge University Press, 1988

Larsen, Henry Asbjorn, *The Northwest Passage 1940–1942 and 1944: the famous voyages of the Royal Canadian Mounted Police Schooner St. Roch*, Ottawa, Queen's Printer, 1958

Lopetegui Torres, Javier, *Antártica – un Desafío Perentorio*, Santiago, Instituto Geopolitico de Chile, 1986

Macfarlane, Stuart, *The Erebus Papers*, Auckland, Avon Press, 1991

Maclellan, David, 'Pullen of the Arctic', *Canadian Geographic*, vol. 104, no. 2, 1984 (pp. 26–33)

Mellor, Malcolm and Charles Swithinbank, 'Airfields on Antarctic glacier ice', *CRREL Report 89–21*, Hanover, New Hampshire, US Army Cold Regions Research and Engineering Laboratory, 1989

Mookhoek, A.D. and W.J. Bielstein, 'Problems associated with the design of an Arctic marine transportation system', *Third Annual Offshore Technology Conference, Houston, Texas, 19–21 April 1971*, 2 (pp. 123–46)

New York Times, 29 September 1969 (p. C25)

Nicholls, K.W., K. Makinson and V.A. Robinson, 'Ocean circulation beneath Ronne Ice Shelf', *Nature*, vol. 354, no. 6350, 1991 (pp. 221–3)

Owen, Roderic, *The Fate of Franklin*, London, Hutchinson, 1978

Pullen, T.C., interviewed in *US News & World Report*, 9 February, 1970 (p. 74)

Reece, Alan, 'The ice of Crown Prince Gustav Channel, Graham Land, Antarctica', *Journal of Glaciology*, vol. 1, no. 8, 1950 (pp. 404–9)

Rees, W.G., 'A new wind-chill nomogram', *Polar Record*, vol. 29, no. 170, 1993 (pp. 229–34)

Reynolds, John M., 'The distribution of mean annual temperatures in the Antarctic Peninsula', *British Antarctic Survey Bulletin*, no. 54, 1981 (pp. 123–33)

Ronne, Finn, 'Ronne Antarctic Research Expedition 1946–48', *Geographical Review*, vol. 38, no. 3, 1948 (pp. 355–91)

Ross, Captain Sir James Clark, *A Voyage of Discovery and Research in the Southern and Antarctic Regions During the Years 1839–43*, 2 vols., London, John Murray, 1847 (vol. 2, p. 332)

Rott, Helmut, Pedro Skvarca and Thomas Nagler, 'Rapid collapse of northern Larsen Ice Shelf, Antarctica', *Science*, vol. 271, no. 5250, 1996 (pp. 788–92)

Rowley, Graham W., 'Captain T.C. Pullen, RCN: Polar Navigator', *The Northern Mariner*, vol. 2, no. 2, 1992 (pp. 29–49)

Rowley, Graham W., *Cold Comfort: My love affair with the Arctic*, Montreal, McGill-Queen's University Press, 1996

Scheina, Robert L., *US Coast Guard Cutters and Craft of World War II*, Annapolis, Maryland, Naval Institute Press (p. 56)

Schytt, Valter, 'Blue ice-fields, moraine features and glacier fluctuations', *Norwegian-British-Swedish Antarctic Expedition 1949–52, Scientific Results*, vol. 4E, Oslo, Norsk Polarinstitutt, 1960

Shackleton, Sir Ernest, *South, the story of Shackleton's last expedition 1914–1917*, London, William Heinemann, 1919

Silverstein, Samuel C., 'The American Antarctic Mountaineering Expedition', *Antarctic Journal of the United States*, vol. 2, no. 2, 1967 (pp. 48–50)

Siple, Paul, *90° South. The Story of the American South Pole conquest*, New York, G.P. Putnam's Sons, 1959 (p. 338)

Smith, B.M. Ewan, 'Airborne radio echo sounding of glaciers in the Antarctic Peninsula', *British Antarctic Survey Scientific Reports*, no. 72, 1972 (p. 8)

Society of Naval Architects and Marine Engineers Transactions, vol. 67, 1959 (p. 184)

Stefansson, Vilhjalmur, *The Northward Course of Empire*, New York, Harcourt Brace, 1922 (pp. 189–99)

Steger, Will and Jon Bowermaster, *Crossing Antarctica*, London, Bantam Press, 1992

Storey, B.C., I.W.D. Dalziel, S.W. Garrett, A.M. Grunow, R.J. Pankhurst and W.R. Vennum, 'West Antarctica in Gondwanaland: Crustal blocks, reconstruction, and breakup processes', *Tectonophysics*, vol. 155, 1988 (pp. 381–90)

Sullivan, Walter, in *The New York Times*, 1 October 1969

Swithinbank, Charles, 'Airborne tourism in the Antarctic', *Polar Record*, vol. 29, no. 169, 1993 (pp. 103–10)

Swithinbank, Charles, 'Arctic pack ice from below', *Sea Ice, Proceedings of an International Conference, Reykjavik, Iceland, May 10–13, 1971*, Reykjavik, National Research Council, 1972 (pp. 246–54)

Swithinbank, Charles, 'Glaciological Research in the Antarctic Peninsula', *Philosophical Transactions of the Royal Society*, vol. B 279, 1977 (pp. 161–83)

Swithinbank, Charles, *Ice Atlas of Arctic Canada*, Ottawa, The Queen's Printer, 1960

Swithinbank, Charles, 'Ice runways in the Heritage Range, Antarctica. A report prepared for Adventure Network International and Fuerza Aérea de Chile', Cambridge, February 1987 (unpublished)

Swithinbank, Charles, 'A new map of Alexander Island, Antarctica', *Polar Record*, vol. 17, no. 107, 1974 (pp. 155–7)

Swithinbank, Charles, *Norwegian-British-Swedish Antarctic Expedition 1949–52, Scientific Results*, vol. 3. Oslo, Norsk Polarinstitutt, 1957–60

Swithinbank, Charles, 'Obituary: John Edward Giles Kershaw', *Polar Record*, vol. 26, no. 158, 1990 (p. 250)

Swithinbank, Charles, 'Radio echo sounding of Antarctic glaciers from light aircraft', *International Association of Scientific Hydrology*, Publication no. 79, 1968 (pp. 405–14)

Swithinbank, Charles, 'Satellite image atlas of glaciers of the world: Antarctica', *US Geological Survey Professional Paper* 1386-B, 1988 (p. 112)

Swithinbank, Charles, SS, Manhattan. 'Notes on a voyage in the Northwest Passage 1969' (proprietary)

Swithinbank, Charles, Christopher Doake, Andrew Wager and Richard Crabtree, 'Major change in the map of Antarctica', *Polar Record*, vol. 18, no. 114, 1976 (pp. 295–9)

Swithinbank, C.W.M. and C. Lane, 'Antarctic Mapping from Satellite Imagery', *Conference of Commonwealth Survey Officers* 1975, chapter 11, London, Ministry of Overseas Development, 1975

Truitt, Joel E., 'Moving Alaska's oil by submarine tanker', *Navy*, vol. 75, no. 3, March 1970 (pp. 80–1)

Tyree, David M., 'New era in the loneliest continent', *National Geographic Magazine*, vol. 123, no. 2, 1963 (p. 283)

Vaughan, D.G. and C.S.M. Doake, 'Recent atmospheric warming and retreat of ice shelves on the Antarctic Peninsula', *Nature*, vol. 379, no. 6563, 1996 (pp. 328–31)

Victor, Paul-Émile, *Mes aventures polaires*, Paris, Éditions G.P., 1975

Walford, M.E.R., 'Glacier movement measured with a radio echo technique', *Nature*, vol. 239, no. 5367, 1972 (pp. 95–6)

Walford, M.E.R., 'Radio echo sounding through an ice shelf', *Nature*, vol. 204, no. 4956, 1964 (pp. 317–19)

Wigley, T.M.L., R. Richels and J.A. Edmonds, 'Economic and environmental choices in the stabilization of atmospheric CO_2 concentrations', *Nature*, vol. 379, no. 6562, 1996 (pp. 240–3)

Williams, Elizabeth, Charles Swithinbank and G. de Q Robin, 'A submarine sonar study of Arctic pack ice', *Journal of Glaciology*, vol. 15, no. 73, 1975 (pp. 349–62)

INDEX

221

224

226

227